SPRINGER PROCEEDINGS IN PHYSICS 101

SPRINGER PROCEEDINGS IN PHYSICS

79 **Nonlinear Dynamics and Pattern Formation in Semiconductors and Devices**
Editor: F.-J. Niedernostheide

80 **Computer Simulation Studies in Condensed-Matter Physics VIII**
Editors: D.P. Landau, K.K. Mon, and H.-B. Schüttler

81 **Materials and Measurements in Molecular Electronics**
Editors: K. Kajimura and S. Kuroda

82 **Computer Simulation Studies in Condensed-Matter Physics IX**
Editors: D.P. Landau, K.K. Mon, and H.-B. Schüttler

83 **Computer Simulation Studies in Condensed-Matter Physics X**
Editors: D.P. Landau, K.K. Mon, and H.-B. Schüttler

84 **Computer Simulation Studies in Condensed-Matter Physics XI**
Editors: D.P. Landau and H.-B. Schüttler

85 **Computer Simulation Studies in Condensed-Matter Physics XII**
Editors: D.P. Landau, S.P. Lewis, and H.-B. Schüttler

86 **Computer Simulation Studies in Condensed-Matter Physics XIII**
Editors: D.P. Landau, S.P. Lewis, and H.-B. Schüttler

87 **Proceedings of the 25th International Conference on the Physics of Semiconductors**
Editors: N. Miura and T. Ando

88 **Starburst Galaxies**
Near and Far
Editors: L. Tacconi and D. Lutz

89 **Computer Simulation Studies in Condensed-Matter Physics XIV**
Editors: D.P. Landau, S.P. Lewis, and H.-B. Schüttler

90 **Computer Simulation Studies in Condensed-Matter Physics XV**
Editors: D.P. Landau, S.P. Lewis, and H.-B. Schüttler

91 **The Dense Interstellar Medium in Galaxies**
Editors: S. Pfalzner, C. Kramer, C. Straubmeier, and A. Heithausen

92 **Beyond the Standard Model 2003**
Editor: H.V. Klapdor-Kleingrothaus

93 **ISSMGE**
Experimental Studies
Editor: T. Schanz

94 **ISSMGE**
Numerical and Theoretical Approaches
Editor: T. Schanz

95 **Computer Simulation Studies in Condensed-Matter Physics XVI**
Editors: D.P. Landau, S.P. Lewis, and H.-B. Schüttler

96 **Electromagnetics in a Complex World**
Editors: I.M. Pinto, V. Galdi, and L.B. Felsen

97 **Fields, Networks, Computational Methods and Systems in Modern Electrodynamics**
A Tribute to Leopold B. Felsen
Editors: P. Russer and M. Mongiardo

98 **Particle Physics and the Universe**
Proceedings of the 9th Adriatic Meeting, Sept. 2003, Dubrovnik
Editors: J. Trampetić and J. Wess

99 **Cosmic Explosions**
On the 10th Anniversary of SN1993J (IAU Colloquium 192)
Editors: J. M. Marcaide and K. W. Weiler

100 **Lasers in the Conservation of Artworks**
LACONA V Proceedings, Osnabrück, Germany, Sept. 15–18, 2003
Editors: K. Dickmann, C. Fotakis, and J.F. Asmus

101 **Progress in Turbulence**
Editors: J. Peinke, A. Kittel, S. Barth, and M. Oberlack

102 **Adaptive Optics for Industry and Medicine**
Proceedings of the 4th International Workshop
Editor: U. Wittrock

Volumes 50–78 are listed at the end of the book.

J. Peinke A. Kittel S. Barth M. Oberlack (Eds.)

Progress in Turbulence

With 111 Figures and 10 Tables

Springer

Professor Dr. Joachim Peinke
PD Dr. habil. Achim Kittel
Dr. Stephan Barth
Carl von Ossietzky Universität, Institute of Physics
Carl-von-Ossietzky-Str. 9–11, 26129 Oldenburg, Germany

Professor Dr.-Ing. Martin Oberlack
TU Darmstadt, Fachbereich für Mechanik und Fachbereich Bauingenieurwesen und Geodäsie
Fachgebiet Strömungsmechanik und Hydraulik
Petersenstr. 13, 64287 Darmstadt, Germany

ISSN 0930-8989

ISBN 3-540-23216-8 Springer Berlin Heidelberg New York

Library of Congress Control Number: 2004117175

Springer is a part of Springer Science+Business Media.

springeronline.com

Printed in The Netherlands

Typesetting and prodcution: PTP-Berlin, Protago-TEX-Production GmbH, Berlin
Cover concept: eStudio Calamar Steinen
Cover production: *design & production* GmbH, Heidelberg

Printed on acid-free paper SPIN: 10973028 89/3141/YU 5 4 3 2 1 0

Preface

Besides turbulence there is hardly any other scientific topic which has been considered as a prominent scientific challenge for such a long time. Although turbulence is based on classical physics and the formulation of the basic equations seems reasonably straightforward, a deep general understanding remains a mystery. To pronounce its importance the Clay Mathematics Institute (Cambridge, Massachusetts) has defined turbulence as one of the seven Millennium problems in Mathematics.

The special interest in turbulence is not only based on it being a difficult scientific problem but also on its meaning in the technical world and our daily life. For example, turbulence has an obvious significance in aeronautics, in the mixing of chemicals in the pharmaceutical industry and even in the daily weather forecast.

This broad interdisciplinary interest has in the past lead to an increasing diversification of the research into distinct disciplines. Thus turbulence is nowadays a topic in mathematics, physics, engineering, geosciences and computer sciences and has lead to different research directions in these fields. In Physics, e.g., the concepts of universality play a central role, whereas the phenomena of real flow situations are partially neglected. For applied engineering problems the famous k-ε model is often employed, although it is known that this approach to turbulence describes physical features of several flow situations incorrectly.

In the last decades turbulence research in the fields of engineering and physics has developed in parallel with a rather poor transfer of knowledge between them. To close the gap between engineering and physical sciences and to improve interchange of knowledge the interdisciplinary turbulence initiative was founded in 1999. In the framework of this initiative about twenty projects each consisting of a cooperation between engineers and physicists are funded by the German science foundation DFG. One important event of the initiative was to organize an international conference which was held in Bad Zwischenahn in September 2003. During the days of the conference an enthusiastic and stimulating atmosphere of discussions and interchange between the different disciplines prevailed.

To capture the different aspects and topics of the meeting the organizers decided to put together a written collection of the contributions and hence make them more accessible to a larger number of scientists. As the editors of this book we decided to pool the contributions together under different headlines to pronounce the different facets of the conference. The topics are mathematical and other approaches to model turbulence, experimental observations, and different methods of simulations. The contributions document the considerable progress made in all fields.

We would like to take the chance to thank all authors for their contributions, the Springer-Verlag, Heidelberg, especially Thomas Ditzinger for his assistance and the DFG for the financial support. Finally we are very grateful to George Khujadze for doing the tedious work of integrating all contributions and giving the book its optical finish.

Oldenburg and Darmstadt
Summer 2004

J. Peinke, M. Oberlack,
A. Kittel and S. Barth,

Contents

Part II Scaling Laws and Intermittency

Part III Modelling

Part I

Mathematical Methods

Passive Scalar Diffusion as a Damped Wave

Axel Brandenburg[1], Petri J. Käpylä[2,3], and Amjed Mohammed[4]

[1] NORDITA, Blegdamsvej 17, 2100 Copenhagen Ø, Denmark,
brandenb@nordita.dk
[2] Kiepenheuer-Institut für Sonnenphysik, Schöneckstr. 6, 79104 Freiburg, Germany
[3] Department of Physical Sciences, Astronomy Division, P.O. Box 3000,
FIN-90014 University of Oulu, Finland, petri.kapyla@oulu.fi
[4] Physics Department, Oldenburg University, 26111 Oldenburg, Germany
amjed@mail.uni-oldenburg.de

Three-dimensional turbulence simulations are used to show that the turbulent root mean square velocity is an upper bound of the speed of turbulent diffusion. There is a close analogy to magnetic diffusion where the maximum diffusion speed is the speed of light. Mathematically, this is caused by the inclusion of the Faraday displacement current which ensures that causality is obeyed. In turbulent diffusion, a term similar to the displacement current emerges quite naturally when the minimal tau approximation is used. Simulations confirm the presence of such a term and give a quantitative measure of its relative importance.

1 Introduction

Since the seminal paper of Prandtl (1925), turbulent diffusion has always been an important application of turbulence theory. By analogy with the kinetic theory of heat conduction, the turbulent exchange of fluid elements leads to an enhanced flux, $\overline{\boldsymbol{\mathcal{F}}}$, of a passive scalar concentration that is proportional to the negative mean concentration gradient,

$$\overline{\boldsymbol{\mathcal{F}}} = -\kappa_{\rm t}\boldsymbol{\nabla}\overline{C} \quad \text{(Fickian diffusion)}, \tag{1}$$

where $\kappa_{\rm t} = \frac{1}{3}u_{\rm rms}\ell_{\rm cor}$ is a turbulent diffusion coefficient, $u_{\rm rms}$ is the turbulent rms velocity, and $\ell_{\rm cor}$ is the correlation length. Equation (1) leads to a closed equation for the evolution of the mean concentration, $\overline{C}$,

$$\frac{\partial\overline{C}}{\partial t} = \kappa_{\rm t}\nabla^2\overline{C}. \tag{2}$$

This is an elliptic equation, which implies that signal propagation is instantaneous and hence causality violating. For example, if the initial $\overline{C}$ profile is a δ-function, it will be a gaussian at the next instant, but gaussians have already infinite support.

The above formalism usually emerges when one considers the microphysics of the turbulent flux in the form $\overline{\boldsymbol{\mathcal{F}}} = \overline{\boldsymbol{u} \int \dot{c}\, \mathrm{d}t}$, where $\dot{c} \approx -\boldsymbol{u} \cdot \boldsymbol{\nabla}\overline{C}$ is the linear approximation to the evolution equation for the fluctuating component of the concentration. Recently, Blackman & Field (2003) proposed that one should instead consider the expression

$$\partial\overline{\boldsymbol{\mathcal{F}}}/\partial t = \overline{\dot{\boldsymbol{u}}c} + \overline{\boldsymbol{u}\dot{c}}. \tag{3}$$

On the right hand side, the nonlinear terms in the two evolution equations for $\boldsymbol{u}$ and c are *not* omitted; they lead to triple correlations which are *assumed* to be proportional to $-\overline{\boldsymbol{\mathcal{F}}}/\tau$, where τ is some relaxation time. Furthermore, there is a priori no reason to omit the time derivative on the left hand side of equation (3). It is this term which leads to the emergence of an extra time derivative (i.e. a 'turbulent displacement flux') in the modified 'non-Fickian' diffusion law,

$$\overline{\boldsymbol{\mathcal{F}}} = -\kappa_{\mathrm{t}}\boldsymbol{\nabla}\overline{C} - \tau\frac{\partial\overline{\boldsymbol{\mathcal{F}}}}{\partial t} \quad \text{(non-Fickian)}. \tag{4}$$

This turns the elliptic equation (2) into a damped wave equation,

$$\frac{\partial^2\overline{C}}{\partial t^2} + \frac{1}{\tau}\frac{\partial\overline{C}}{\partial t} = \tfrac{1}{3}u_{\mathrm{rms}}^2\nabla^2\overline{C}. \tag{5}$$

The maximum wave speed is obviously $u_{\mathrm{rms}}/\sqrt{3}$. Note also that, after multiplication with τ, the coefficient on the right hand side becomes $\frac{1}{3}\tau u_{\mathrm{rms}}^2 = \kappa_{\mathrm{t}}$, and the second time derivative on the left hand side becomes unimportant in the limit $\tau \to 0$, or when the physical time scales are long compared with τ.

2 Validity of Turbulent Displacement Flux and Value of τ

A particularly obvious way of demonstrating the presence of the second time derivative is by considering a numerical experiment where $\overline{C} = 0$ initially. Equation (2) would predict that then $\overline{C} = 0$ at all times. But, according to the alternative formulation (5), this need not be true if initially $\partial\overline{C}/\partial t \neq 0$. In practice, this can be achieved by arranging the initial fluctuations of c such that they correlate with u_z. Of course, such highly correlated arrangement will soon disappear and hence there will be no turbulent flux in the long time limit. Nevertheless, at early times, $\langle\overline{C}^2\rangle^{1/2}$ (a measure of the passive scalar amplitude) rises from zero to a finite value; see Fig. 1.

Closer inspection of Fig. 1 reveals that when the wavenumber of the forcing is sufficiently small (i.e. the size of the turbulent eddies is comparable to the

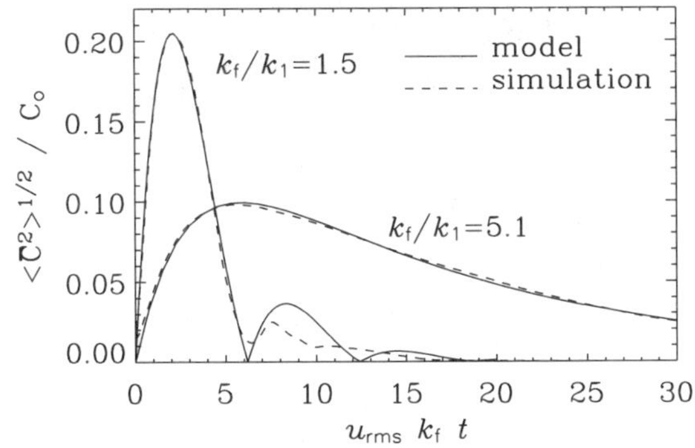

Fig. 1. Passive scalar amplitude, $\langle\overline{C}^2\rangle^{1/2}$, versus time (normalized by $u_{\rm rms}k_{\rm f}$) for two different values of $k_{\rm f}/k_1$. The simulations have 256^3 meshpoints. The results are compared with solutions to the non-Fickian diffusion model.

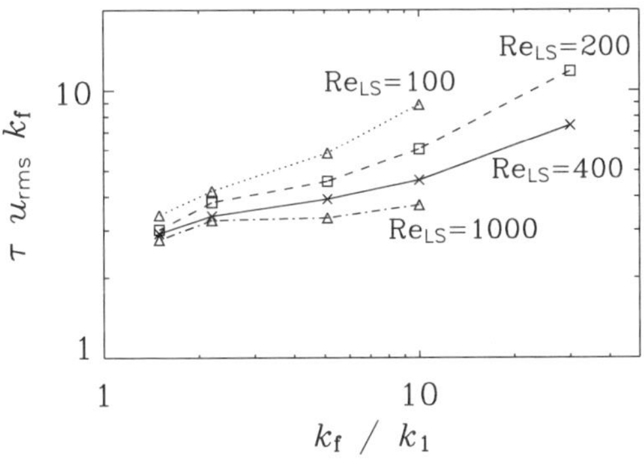

Fig. 2. Strouhal number as a function of $k_{\rm f}/k_1$ for different values of $\mathrm{Re}_{\rm LS}$, i.e. the large scale Reynolds number. The resolution varies between 64^3 meshpoints ($\mathrm{Re}_{\rm LS}=100$) and 512^3 meshpoints ($\mathrm{Re}_{\rm LS}=1000$).

box size), $\langle\overline{C}^2\rangle^{1/2}$ approaches zero in an oscillatory fashion. This remarkable result can only be explained by the presence of the second time derivative term giving rise to wave-like behavior. This shows that the presence of the new term is actually justified. Comparison with model calculations shows that the non-dimensional measure of τ, $\mathrm{St}\equiv\tau u_{\rm rms}k_{\rm f}$, must be around 3. (In mean-field theory this number is usually called Strouhal number.) This rules out the validity of the quasilinear (first order smoothing) approximation which would only be valid for $\mathrm{St}\to 0$.

Next, we consider an experiment to establish directly the value of St. We do this by imposing a passive scalar gradient, which leads to a steady state, and

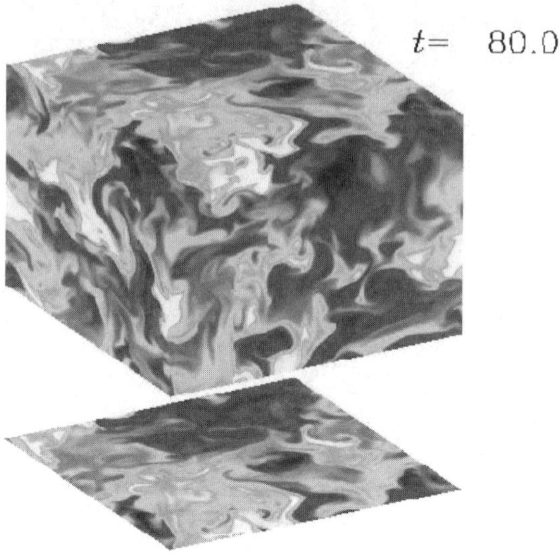

Fig. 3. Visualizations of C on the periphery of the simulation domain at a time when the simulation has reached a statistically steady state. $k_f/k_1 = 1.5$, $\mathrm{Re}_{LS} = 400$.

measuring the resulting turbulent passive scalar flux. By comparing double and triple moments we can measure St quite accurately without invoking a fitting procedure as in the previous experiment. The result is shown in Fig. 2 and confirms that $\mathrm{St} \approx 3$ in the limit of small forcing wavenumber, k_f. The details can be found in Brandenburg et al. (2004). A Visualization of C on the periphery of the simulation domain is shown in Fig. 3 for $k_f = 1.5$. Note the combination of large patches (scale $\sim 1/k_f$) together with thin filamentary structures.

Finally, we should note that equation (3) in the passive scalar problem was originally motivated by a corresponding expression for the electromotive force in dynamo theory, where the $\dot{\boldsymbol{u}}$ terms leads to the crucial nonlinearity of the α-effect (Blackman & Field 2002).

References

1. Blackman, EG, Field, GB (2002) New dynamical mean-field dynamo theory and closure approach. Phys. Rev. Lett. 89:265007
2. Blackman, EG, Field, GB (2003) A simple mean field approach to turbulent transport. Phys. Fluids 15:L73–L76
3. Brandenburg, A, Käpylä, P, & Mohammed, A (2004) Non-Fickian diffusion and tau-approximation from numerical turbulence Phys. Fluids (in press)
4. Prandtl, L (1925) Bericht über Untersuchungen zur ausgebildeten Turbulenz. Zeitschr. Angewandt. Math. Mech.. 5:136–139

Extremalizing Vector Fields as Guides Toward Understanding Properties of Turbulence

F.H. Busse

Institute of Physics, University of Bayreuth, D-95440 Bayreuth
busse@uni-bayreuth.de

1 Introduction

The theory of rigorous upper bounds on turbulent transports has just passed its fortieth anniversary since it was in 1963 that Howard's pioneering paper has been published. The theory has progressed considerably since that time and may still not have reached its full potential. Howard's analysis[1] provides upper bounds on the heat transport by turbulent convection in a fluid layer heated from below and its methods have later been applied to problems of transports of mass, linear momentum and angular momentum [2, 3, 4, 5]. A significant progress was achieved through the discovery of the multi-alpha-solutions of the variational problems in the case when the equation of continuity is imposed as constraint on the extremalizing vector fields [6]. In the 1990ies the theory of upper bounds experienced a revival through the independent development of the background field method [7, 8]. Although the background field is initially introduced as a mathematical device to facilitate the derivation of bounds for various terms in the energy balance, when optimized the method leads to the same bounds on turbulent transports as the Howard-Busse method of solving Euler-Lagrange equations [9, 10, 11]. The equivalence of the two methods has been demonstrated in an important paper by Kerswell [12].

While rigorous bounds on global properties of turbulent states of fluid flow are welcome results, certain features of the extremalizing vector fields may be of even deeper interest because of similarities with properties of the realized turbulent velocity fields. In order to discuss these aspects of the optimum theory of turbulence, - as the upper bound theory is sometimes called -, we first have to outline its procedure. Thereafter we shall discuss the ways in which solutions of the variational problems may help to understand properties of turbulent flows and relationships between different turbulent systems. An example for the latter is relationship between the turbulent transport of momentum and the turbulent resistivity in a layer of an electrically conducting fluid that has recently been studied [13].

2 Formulation of the Variational Problem for Turbulent Couette Flow

In order to introduce the optimum theory of turbulence we consider the simplest configuration for which a non-trivial solution of the Navier-Stokes equations of motion exists: the configuration of plane Couette flow. The Reynolds number is defined in this case in terms of the constant relative motion $U_0\mathbf{i}$ between the plates, $Re = U_0 d/\nu$, where $\mathbf{i}$ is the unit vector parallel to the plates and ν is the kinematic viscosity of the fluid. Using the distance d between the plates as length scale and d^2/ν as time scale the basic equations can be written in dimensionless form,

$$\frac{\partial}{\partial t}\mathbf{v} + \mathbf{v}\cdot\nabla\mathbf{v} = -\nabla p + \nabla^2\mathbf{v}, \tag{1}$$

$$\nabla\cdot\mathbf{v} = 0. \tag{2}$$

We use a cartesian system of coordinates with the x, z-coordinates in the directions of $\mathbf{i}$ and $\mathbf{k}$, respectively, where $\mathbf{k}$ is the unit vector normal to the plates such that the boundary conditions are given by

$$\mathbf{v} = \mp\frac{1}{2}Re\,\mathbf{i} \text{ at } z = \pm\frac{1}{2}. \tag{3}$$

After separating the velocity field $\mathbf{v}$ into its mean and fluctuating parts, $\mathbf{v} = \mathbf{U} + \check{\mathbf{v}}$ with $\overline{\mathbf{v}} = \mathbf{U}$, $\overline{\check{\mathbf{v}}} = 0$, where the bar denotes the average over planes $z = const.$ we obtain by multiplying equation (1) by $\check{\mathbf{v}}$ and averaging it over the entire fluid layer (indicated by angular brackets)

$$\frac{1}{2}\frac{d}{dt}\langle|\,\check{\mathbf{v}}\,|^2\rangle = -\langle\overline{\mathbf{u}w}\cdot\frac{\partial}{\partial z}\mathbf{U}\rangle - \langle|\,\nabla\check{\mathbf{v}}\,|^2\rangle. \tag{4}$$

Here $\mathbf{u}$ denotes the component of $\check{\mathbf{v}}$ perpendicular to $\mathbf{k}$ and w is its $z-$ component. If we define fluid turbulence under stationary conditions by the property that quantities averaged over planes $z = const.$ are time independent, the equation for the mean flow $\mathbf{U}$ can be integrated to yield

$$\frac{d}{dz}\mathbf{U} = \overline{w\mathbf{u}} - \langle w\mathbf{u}\rangle - Re\,\mathbf{i} \tag{5}$$

where the boundary condition (3) has been employed. With this relationship $\mathbf{U}$ can be eliminated from the problem and the energy balance

$$\langle|\,\nabla\mathbf{u}\,|^2\rangle + \langle|\,\overline{\mathbf{u}w} - \langle\mathbf{u}w\rangle\,|^2\rangle = Re\langle u_x w\rangle \tag{6}$$

is obtained where the identity $\langle\overline{\mathbf{u}w}^2\rangle - \langle\mathbf{u}w\rangle^2 = \langle|\overline{\mathbf{u}w} - \langle\mathbf{u}w\rangle|^2\rangle$ has been used.

Since the momentum transport in the $x-$direction between the moving rigid plates is described by $M = -dU_x/dz\,|_{z=\pm\frac{1}{2}} = \langle u_x w\rangle + Re$ we can conclude immediately that the momentum transport by turbulent flow always exceeds the corresponding laminar value because $\langle u_x w \rangle$ is positive according to relationship (6). Since a lower bound on M thus exists an upper bound μ on $\langle u_x w \rangle$ as a function of Re is of primary interest. Following Howard [1] it can be shown that $\mu(Re)$ is a monotonous function and it is therefor equivalent to ask for a lower bound R of Re at a given value μ of $\langle u_x w \rangle$. We are thus led to the following formulation of the variational problem:

Find the minimum $R(\mu)$ of the functional

$$\mathcal{R}(\tilde{\mathbf{v}}, \mu) \equiv \frac{\langle | \nabla \tilde{\mathbf{v}} |^2 \rangle}{\langle \tilde{u}_x \tilde{w} \rangle} + \mu \frac{\langle | \overline{\tilde{\mathbf{u}}\tilde{w}} - \langle \tilde{\mathbf{u}}\tilde{w} \rangle |^2 \rangle}{\langle \tilde{u}_x \tilde{w} \rangle^2} \tag{7}$$

among all solenoidal vector fields $\tilde{\mathbf{v}} = \tilde{\mathbf{u}} + \mathbf{k}\tilde{w}$ that satisfy the boundary condition $\tilde{\mathbf{v}} = 0$ at $z = \pm\frac{1}{2}$ and the condition $\langle \tilde{u}_x \tilde{w} \rangle > 0$.

The Euler–Lagrange equations as necessary conditions for an extremal value of the functional are given by

$$\tilde{w} \frac{d}{dz} \mathbf{U}^* + \mathbf{k}\, \tilde{\mathbf{u}} \cdot \frac{d}{dz} \mathbf{U}^* = -\nabla \pi + \nabla^2 \tilde{\mathbf{v}}, \tag{8}$$

$$\nabla \cdot \tilde{\mathbf{v}} = 0, \tag{9}$$

where $\frac{d}{dz}\mathbf{U}^*$ is defined by

$$\frac{d}{dz}\mathbf{U}^* = \overline{\tilde{\mathbf{u}}\tilde{w}} - \langle \tilde{\mathbf{u}}\tilde{w} \rangle - \mathbf{i}\left(R - \frac{\langle | \nabla \tilde{\mathbf{v}} |^2 \rangle}{2\langle \tilde{u}_x \tilde{w} \rangle}\right). \tag{10}$$

When equations (8),(9),(10) are compared with the equations for $\check{\mathbf{v}}$ and for $\mathbf{U}$ a strong similarity can be noticed. The variational problem does not exhibit any time dependence, but the Euler–Lagrange equations may still be regarded as the symmetric analogue of the Navier-Stokes equations for steady flow.

3 Properties of the Extremalizing Vector Fields

For lack of space this section must be kept brief and the reader is referred to the review articles [14, 15] for details. The n-alpha-solutions of equations (8),(9),(10) are characterized by a hierarchy of n boundary layers at each plate and provide the upper bound sequentially with increasing R starting with n=1. The extremalizing vector fields thus exhibit a bifurcation structure similar to that found in many cases of the transition to turbulence. The thicknesses of the boundary layers decrease with increasing R and their ratio from one layer to the next approaches the factor 4. The typical scale of motion thus increases linearly with distance from the wall as assumed in Prandtl's mixing length theory. But the discreteness of the scales reflects the fact that effective transports require preferred scales. Cloud patterns in the atmosphere and distinct scales of convection on the sun remind us that highly turbulent systems do indeed exhibit spontaneous discrete scales.

Of particular interest is the mean flow $\mathbf{U}^*$ which exhibits a finite shear of the order $-Re/4$ in the interior. It has been a puzzle that a finite shear is also observed in experiments on turbulent Couette flow [16] while in the case of turbulent convection in a layer heated from below the mean temperature becomes isothermal in the interior in experiments as well as for the extremaling vector field. It appears that only part of the available imposed shear is needed for the boundary layers in order to achieve a maximum transport of momentum. Another case of good agreement between the profile $\mathbf{U}^*$ and measurements can be found in the case of turbulent flow between differentially rotating coaxial cylinders [17, 15].

The close similarities between variational functionals for rather different physical systems suggest analogous similarities between the respective turbulent fields. The analogy between the fluctuating component of the temperature in turbulent convection and the streamwise component of the fluctuating velocity field in shear flow turbulence has been demonstrated in [4] and has been applied in the theory of atmospheric boundary layer [18]. More properties of the extremalizing vector fields become available for comparisons when additional constraints are imposed. For example, the energy balances for poloidal and toroidal components of the velocity field can be applied separately [19].

References

1. Howard, L.N. (1963) *J. Fluid Mech.* 17:405–432
2. Busse, F.H. (1968) *Z. Angew. Math. Mech.* 48:T187–T190.
3. Busse, F.H. (1969a) *J. Appl. Math. Phys. (ZAMP)* 20:1–14
4. Busse, F.H. (1970) *J. Fluid Mech.* 41:219–240
5. Busse, F.H. (1972) *Springer Lect. Notes Phys.* 12:103–126
6. Busse, F.H. (1969b) *J. Fluid Mech.* 37:457–477
7. Doering, C.R., and Constantin, P. (1992) *Phys. Rev. Lett.* 69:1608–1651
8. Doering, C.R., and Constantin, P. (1994) *Phys. Rev. E* 49:4087–4099
9. Nicodemus, R., Grossmann, S., and Holthaus, M. (1997) *PhysicaD*101:178–190
10. Nicodemus, R., Grossmann, S., and Holthaus, M. (1997) *Phys. Rev. E* 56:6774–6786.
11. Plasting, S.C., and Kerswell, R.R. (2003) *J. Fluid Mech.* 477:363–379
12. Kerswell, R.R. (1998) *PhysicaD* 121:175–192
13. Busse, F.H. (2004) *J. Plasma Physics* 70:1–7
14. Busse, F.H. (1978) *Advances in Appl. Mech.* 18:77-121.
15. Busse, F. H. (2002) The Problem of Turbulence and the Manifold of Asymptotic Solutions of the Navier-Stokes Equations, pp. 77-121 in *Theories of Turbulence* M. Oberlack and F. H. Busse (eds) Springer, Wien, New York
16. Reichardt, H. (1959) Gesetzmäßigkeiten der geradlinigen turbulenten Couetteströmung *Mitt. Max-Planck-Institut f. Strömungsforsch., Göttingen, Nr. 22*
17. Busse, F.H. (1996) *Springer Lect. Notes Phys.* 476:1–9
18. Deardorff, J.W. (1970) *J. Atmos. Sci.* 27:1211–1213
19. Vitanov, N.K., and Busse, F.H. (2001) *Phys. Rev. E* 63:16303–16310.

Low-Wavenumber Forcing and Turbulent Energy Dissipation

Charles R. Doering and Nikola P. Petrov

Department of Mathematics and Michigan Center for Theoretical Physics
University of Michigan, Ann Arbor, MI 48109, USA
doering@umich.edu and npetrov@umich.edu

1 Introduction

In many Direct Numerical Simulations (DNS) of turbulence researchers inject power into the fluid at large scales and then observe how it "propagates" to the small scales [1, 2, 3, 4, 5, 6, 7, 8, 9, 10, 11, 12]. One such type of stirring is to take the force $\mathbf{f}(\mathbf{x}, t)$ to be proportional to the projection of the velocity $\mathbf{u}(\mathbf{x}, t)$ of the flow onto its lowest Fourier modes, while keeping the rate of injected external power constant. In this paper we perform a simple but rigorous analysis to establish bounds on the relationship between the energy dissipation rate (which is the same as the injected power) and the resulting Reynolds number. While this analysis cannot give detailed information of the energy spectrum, it does provide some indication of the balance of energy between the lower, directly forced, modes, and those excited by the cascade. This work is an extension of the analysis in [13, 14, 15], where the force is fixed (not a functional of the velocity).

Consider fluid in a periodic d-dimensional box of side length ℓ. The allowed wave vectors $\mathbf{k}$ are of the form $\mathbf{k} = \frac{2\pi}{\ell}\mathbf{a}$, where $\mathbf{a} \in \mathbb{Z}^d$ is a d-dimensional vector with integer components. Let $\mathcal{L}$ be the subset of wave vectors that have the smallest possible wavenumber (namely, $\frac{2\pi}{\ell}$); $\mathcal{L}$ consists of $2d$ elements: $\mathcal{L} = \{\pm\frac{2\pi}{\ell}\mathbf{e}_1, \ldots, \pm\frac{2\pi}{\ell}\mathbf{e}_d\}$. The operator $\mathcal{P}$ projects the vector field

$$\mathbf{u}(\mathbf{x}, t) = \sum_{\mathbf{k}} \hat{\mathbf{u}}(\mathbf{k}, t)\, \mathrm{e}^{\mathrm{i}\mathbf{k}\cdot\mathbf{x}}$$

onto the subspace spanned by the Fourier components with wave vectors in $\mathcal{L}$:

$$\mathcal{P}\mathbf{u}(\mathbf{x}, t) = \sum_{\mathbf{k}\in\mathcal{L}} \hat{\mathbf{u}}(\mathbf{k}, t)\, \mathrm{e}^{\mathrm{i}\mathbf{k}\cdot\mathbf{x}} . \tag{1}$$

Obviously, $\mathcal{P}$ maps L^2 into L^2 vector fields; in fact, $\mathcal{P}\mathbf{u}$ is C^∞ in the spatial variables. The projection also preserves the incompressibility property. That is, if $\nabla \cdot \mathbf{u}(\mathbf{x}, t) = 0$, then $\nabla \cdot \mathcal{P}\mathbf{u}(\mathbf{x}, t) = 0$.

The Navier-Stokes equation is

$$\dot{\mathbf{u}} + (\mathbf{u} \cdot \nabla)\mathbf{u} + \frac{1}{\rho}\nabla p = \nu \Delta \mathbf{u} + \mathbf{f} \ , \tag{2}$$

with $\mathbf{f}(\mathbf{x}, t)$ taken in the form

$$\mathbf{f}(\mathbf{x}, t) = \epsilon \, \frac{\mathcal{P}\mathbf{u}(\mathbf{x}, t)}{\frac{1}{\ell^d} \|\mathcal{P}\mathbf{u}(\cdot, t)\|_2^2} \ . \tag{3}$$

where $\|\cdot\|_2$ stands for the L^2-norm, $\|\mathcal{P}\mathbf{u}(\cdot, t)\|_2 := \left[\int |\mathcal{P}\mathbf{u}(\mathbf{x}, t)|^2 \, \mathrm{d}^d\mathbf{x}\right]^{\frac{1}{2}}$.
This choice of forcing ensures that the input power is constant:

$$\int \mathbf{u}(\mathbf{x}, t) \cdot \mathbf{f}(\mathbf{x}, t) \, \mathrm{d}^d\mathbf{x} = \ell^d \epsilon \ . \tag{4}$$

In this approach ϵ, ν and ℓ are the (only) control parameters. On average, the power input is the viscous energy dissipation rate:

$$\epsilon := \frac{1}{\ell^d} \int \mathbf{u}(\mathbf{x}, t) \cdot \mathbf{f}(\mathbf{x}, t) \, \mathrm{d}^d\mathbf{x} = \nu \, \frac{\langle \|\nabla \mathbf{u}\|_2^2 \rangle}{\ell^d} \ , \tag{5}$$

where $\langle \cdot \rangle$ stands for the long time average. The non-dimensional measure of energy dissipation is defined as

$$\beta := \frac{\epsilon \ell}{U^3} \ , \tag{6}$$

which is a function of $\mathsf{Re} := \frac{U\ell}{\nu}$, the Reynolds number, where U is the r.m.s. velocity defined by $U^2 := \frac{\langle \|\mathbf{u}\|_2^2 \rangle}{\ell^d}$, a measure of the total kinetic energy of the fluid. Our analysis will establish limits on the relationship between β and Re.

Because we will study the "low-k" Fourier modes (i.e., modes with wave vectors in $\mathcal{L}$), we also introduce the r.m.s. velocity V contained in these modes,

$$V^2 := \frac{\langle \|\mathcal{P}\mathbf{u}\|_2^2 \rangle}{\ell^d} \ . \tag{7}$$

The bounds on the dissipation β will be in terms of Re and the quantity

$$\mathfrak{p} := \frac{V}{U} \sim \sqrt{\frac{\text{"low-}k\text{" kinetic energy of the fluid}}{\text{Total kinetic energy of the fluid}}} \ . \tag{8}$$

The case $\mathfrak{p} \approx 1$ corresponds to laminar flow, when the turbulent cascade is inoperative.

2 Derivation of the Bounds

2.1 Lower Bounds on the Energy Dissipation

To obtain lower bounds on the energy dissipation, we proceed as usual by multiplying the Navier-Stokes equation (2) by $\mathbf{u}(\mathbf{x},t)$ and integrating over the volume of the fluid to obtain the instantaneous power balance,

$$\frac{1}{2}\frac{\mathrm{d}}{\mathrm{d}t}\,\|\mathbf{u}(\cdot,t)\|_2^2 = -\nu\,\|\nabla\mathbf{u}(\cdot,t)\|_2^2 + \ell^d\epsilon\;, \tag{9}$$

where $\|\nabla\mathbf{u}(\cdot,t)\|_2^2 := \int \left|\sum_{j,m=1}^{d} \partial_j u_m(\mathbf{x},t)\right|^2 \mathrm{d}^d\mathbf{x}$.

Now we use the facts that the lengths of wavevectors $\mathbf{k} \notin \mathcal{L}$ are at least $2\pi\sqrt{2}/\ell$, and that $\|\mathbf{u}(\cdot,t) - \mathcal{P}\mathbf{u}(\cdot,t)\|_2^2 = \|\mathbf{u}(\cdot,t)\|_2^2 - \|\mathcal{P}\mathbf{u}(\cdot,t)\|_2^2$, to derive a lower bound on $\|\nabla\mathbf{u}(\cdot,t)\|_2^2$:

$$\begin{aligned}
\|\nabla\mathbf{u}(\cdot,t)\|_2^2 &= \int |\nabla\mathbf{u}(\mathbf{x},t)|^2\,\mathrm{d}^d\mathbf{x} = \ell^d\sum_{\mathbf{k}} k^2|\hat{\mathbf{u}}(\mathbf{k},t)|^2 \\
&= \ell^d\left(\sum_{\mathbf{k}\in\mathcal{L}} k^2|\hat{\mathbf{u}}(\mathbf{k},t)|^2 + \sum_{\mathbf{k}\notin\mathcal{L}} k^2|\hat{\mathbf{u}}(\mathbf{k},t)|^2\right) \\
&\geq \ell^d\,\frac{4\pi^2}{\ell^2}\left(\sum_{\mathbf{k}\in\mathcal{L}} |\hat{\mathbf{u}}(\mathbf{k},t)|^2 + 2\sum_{\mathbf{k}\notin\mathcal{L}} |\hat{\mathbf{u}}(\mathbf{k},t)|^2\right) \\
&= \frac{4\pi^2}{\ell^2}\left(\|\mathcal{P}\mathbf{u}(\cdot,t)\|_2^2 + 2\,\|\mathbf{u}(\cdot,t) - \mathcal{P}\mathbf{u}(\cdot,t)\|_2^2\right) \\
&= \frac{4\pi^2}{\ell^2}\left(2\,\|\mathbf{u}(\cdot,t)\|_2^2 - \|\mathcal{P}\mathbf{u}(\cdot,t)\|_2^2\right)\;.
\end{aligned} \tag{10}$$

From (9) and (10) we obtain the differential inequality

$$\frac{1}{2}\frac{\mathrm{d}}{\mathrm{d}t}\,\|\mathbf{u}(\cdot,t)\|_2^2 \leq -\nu\,\frac{4\pi^2}{\ell^2}\,\|\mathbf{u}(\cdot,t)\|_2^2 + \ell^d\epsilon\;,$$

from which, using Gronwall's inequality, we deduce

$$\frac{1}{2}\,\|\mathbf{u}(\cdot,t)\|_2^2 \leq \frac{1}{2}\,\|\mathbf{u}(\cdot,0)\|_2^2\,\mathrm{e}^{-\frac{8\pi^2\nu}{\ell^2}t} + \ell^d\epsilon\frac{\ell^2}{8\pi^2\nu}\left(1 - \mathrm{e}^{-\frac{8\pi^2\nu}{\ell^2}t}\right)\;. \tag{11}$$

The inequality (11) implies that $\|\mathbf{u}(\cdot,t)\|_2^2$ is bounded uniformly in time, which in turn implies that the time average of the time derivative in (9) vanishes. This ensures that the time-averaged power balance (assuming that the limit associated with the long time average exists) is indeed given by (5).

Taking the time average of (10), we obtain the bound

$$\frac{4\pi^2\nu}{\ell^2}\left(2U^2 - V^2\right) \le \epsilon \ ,$$

which in non-dimensional variables reads

$$\frac{4\pi^2}{\mathsf{Re}}(2-\mathsf{p}^2) \le \beta \ . \tag{12}$$

2.2 Upper Bound on the Energy Dissipation

To derive an upper bound on β, we multiply the Navier-Stokes equation (2) by $\frac{\mathcal{P}\mathbf{u}(\mathbf{x},t)}{\|\mathcal{P}\mathbf{u}(\cdot,t)\|_2}$ and integrate. The term with $\dot{\mathbf{u}}$ gives a total time derivative,

$$\int \dot{\mathbf{u}}\cdot\frac{\mathcal{P}\mathbf{u}}{\|\mathcal{P}\mathbf{u}\|_2}\,\mathrm{d}^d\mathbf{x} = \frac{1}{\|\mathcal{P}\mathbf{u}\|_2}\int\frac{\partial}{\partial t}(\mathcal{P}\mathbf{u})\cdot\mathcal{P}\mathbf{u}\,\mathrm{d}^d\mathbf{x} = \frac{1}{2}\frac{\mathrm{d}\,\|\mathcal{P}\mathbf{u}(\cdot,t)\|_2}{\mathrm{d}t}\ .$$

For the viscosity term we obtain (integrating by parts)

$$\nu\int(\Delta\mathbf{u})\cdot\frac{\mathcal{P}\mathbf{u}}{\|\mathcal{P}\mathbf{u}\|_2}\,\mathrm{d}^d\mathbf{x} = -\nu\,\frac{\|\nabla\mathcal{P}\mathbf{u}\|_2^2}{\|\mathcal{P}\mathbf{u}\|_2} = -\nu\,\frac{4\pi^2}{\ell^2}\,\|\mathcal{P}\mathbf{u}\|_2\ ,$$

while the forcing term gives $\ell^d\epsilon/\|\mathcal{P}\mathbf{u}(\cdot,t)\|_2$ (cf. (4)).

To estimate the inertial term, we introduce temporarily the notation $\mathbf{p}(\mathbf{x},t) := \mathcal{P}\mathbf{u}(\mathbf{x},t)$. We will make use of the uniform (in $\mathbf{x}$ and t) estimate

$$\frac{|\partial_j p_m(\mathbf{x},t)|}{\|\mathbf{p}(\cdot,t)\|_2} \le \sum_{\mathbf{k}\in\mathcal{L}}\frac{|k_j|\,|\hat{u}_m(\mathbf{k},t)|}{\|\mathbf{p}(\cdot,t)\|_2} \le \frac{2\pi}{\ell^{1+\frac{d}{2}}}\,\frac{\sum_{\mathbf{k}\in\mathcal{L}}|\hat{\mathbf{u}}(\mathbf{k},t)|}{\sqrt{\sum_{\mathbf{k}'\in\mathcal{L}}|\hat{\mathbf{u}}(\mathbf{k}',t)|^2}} \le \frac{2\pi\sqrt{d}}{\ell^{1+\frac{d}{2}}}\ . \tag{13}$$

Then the inertial term may be bounded (we use $\nabla\cdot\mathbf{p}=0$):

$$\begin{aligned}\left|\int[(\mathbf{u}\cdot\nabla)\mathbf{u}]\cdot\frac{\mathbf{p}}{\|\mathbf{p}\|_2}\,\mathrm{d}^d\mathbf{x}\right| &= \left|\int\mathbf{u}\cdot\frac{\nabla\mathbf{p}}{\|\mathbf{p}\|_2}\cdot\mathbf{u}\,\mathrm{d}^d\mathbf{x}\right| \\ &\le \frac{2\pi\sqrt{d}}{\ell^{1+(d/2)}}\int|\mathbf{u}|^2\,\mathrm{d}^d\mathbf{x} = \frac{2\pi\sqrt{d}}{\ell^{1+(d/2)}}\,\|\mathbf{u}\|_2^2\ . \end{aligned}\tag{14}$$

This estimate, however, is obviously not going to be tight for small Re, when the flow is not very turbulent. To improve this estimate so that it take into account the fact that for small Re the energy does not "propagate" much from the large to the small wavenumbers, we split the velocity $\mathbf{u}$ into a "low-k" component, $\mathcal{P}\mathbf{u}$, and a "high-k" one, $\mathbf{u}-\mathcal{P}\mathbf{u}$. We will still use the uniform estimate (13) as well as the inequality $ab \le \frac{1}{2}(za^2+\frac{1}{z}b^2)$ which holds for any $z>0$:

$$\left|\int [(\mathbf{u}\cdot\nabla)\mathbf{u}]\cdot\frac{\mathbf{p}}{\|\mathbf{p}\|_2}\,\mathrm{d}^d\mathbf{x}\right| = \left|\int [\mathbf{p}+(\mathbf{u}-\mathbf{p})]\cdot\frac{\nabla\mathbf{p}}{\|\mathbf{p}\|_2}\cdot[\mathbf{p}+(\mathbf{u}-\mathbf{p})]\,\mathrm{d}^d\mathbf{x}\right|$$

$$\le \frac{2\pi\sqrt{d}}{\ell^{1+(d/2)}}\int \left(2\,|\mathbf{p}|\,|\mathbf{u}-\mathbf{p}| + |\mathbf{u}-\mathbf{p}|^2\right)\mathrm{d}^d\mathbf{x}$$

$$\le \frac{2\pi\sqrt{d}}{\ell^{1+(d/2)}}\int \left[z|\mathbf{p}|^2 + \left(\tfrac{1}{z}+1\right)|\mathbf{u}-\mathbf{p}|^2\right]\mathrm{d}^d\mathbf{x}$$

$$\le \frac{2\pi\sqrt{d}}{\ell^{1+(d/2)}}\left[\left(\tfrac{1}{z}+1\right)\|\mathbf{u}\|_2^2 + \left(z-\tfrac{1}{z}-1\right)\|\mathbf{p}\|_2^2\right] . \tag{15}$$

Putting together (14) and (15), we find

$$\ell^d\epsilon\frac{1}{\|\mathcal{P}\mathbf{u}(\cdot,t)\|_2} \le \frac{1}{2}\frac{\mathrm{d}\,\|\mathcal{P}\mathbf{u}(\cdot,t)\|_2}{\mathrm{d}t}$$

$$+\frac{2\pi\sqrt{d}}{\ell^{1+(d/2)}}\min\left\{\|\mathbf{u}\|_2^2,\ \left(\tfrac{1}{z}+1\right)\|\mathbf{u}\|_2^2 + \left(z-\tfrac{1}{z}-1\right)\|\mathcal{P}\mathbf{u}\|_2^2\right\}$$

$$+\,\nu\,\frac{4\pi^2}{\ell^2}\,\|\mathcal{P}\mathbf{u}(\cdot,t)\|_2 . \tag{16}$$

Now take the time average of all terms in the above inequality. First note that the average of the time derivative of $\|\mathcal{P}\mathbf{u}(\cdot,t)\|_2$ gives zero due to the boundedness of $\|\mathcal{P}\mathbf{u}(\cdot,t)\|_2$ (which follows from the boundedness of $\|\mathbf{u}(\cdot,t)\|_2$; see (11)). To estimate the other terms, we use Jensen's inequality: if a function θ is convex and $\langle\cdot\rangle$ stands for averaging, then $\langle\theta\circ g\rangle \ge \theta\left(\langle g\rangle\right)$ for any real-valued function g. Applying this inequality to the case $g(t) = \|\mathcal{P}\mathbf{u}(\cdot,t)\|_2$ and the convex function $\theta(t) = t^2$, we obtain (same as Cauchy-Schwarz)

$$\langle\|\mathcal{P}\mathbf{u}\|_2\rangle \le \sqrt{\langle\|\mathcal{P}\mathbf{u}\|_2^2\rangle} = \ell^{d/2}\,V .$$

On the other hand, if we take $\theta(t) = \frac{1}{t}$ for $t > 0$, we deduce

$$\left\langle\frac{1}{\|\mathcal{P}\mathbf{u}\|_2}\right\rangle \ge \frac{1}{\langle\|\mathcal{P}\mathbf{u}\|_2\rangle} \ge \frac{1}{\sqrt{\langle\|\mathcal{P}\mathbf{u}\|_2^2\rangle}} = \frac{1}{\ell^{d/2}\,V} .$$

Plugging these estimates into (16), we obtain

$$\epsilon \le \nu\frac{4\pi^2}{\ell^2}V^2 + \frac{2\pi\sqrt{d}}{\ell}\min\left\{U^2V,\ \left(\tfrac{1}{z}+1\right)U^2V + \left(z-\tfrac{1}{z}-1\right)V^3\right\} .$$

In terms of the non-dimensional energy dissipation rate (6), we can rewrite this inequality in the form

$$\beta \le \frac{4\pi^2}{\mathsf{Re}}\,\mathfrak{p}^2 + 2\pi\sqrt{d}\,\phi(\mathfrak{p},z) , \tag{17}$$

where we have introduced the function

$$\phi(\mathfrak{p},z) := \min\left\{\mathfrak{p},\ \left(\tfrac{1}{z}+1\right)\mathfrak{p} + \left(z-\tfrac{1}{z}-1\right)\mathfrak{p}^3\right\} . \tag{18}$$

2.3 Compatibility of the Lower and Upper Bounds on β

Assembling the lower and upper bounds (12) and (17), we have

$$\frac{4\pi^2}{\mathsf{Re}}(2-\mathfrak{p}^2) \le \beta \le \frac{4\pi^2}{\mathsf{Re}}\,\mathfrak{p}^2 + 2\pi\sqrt{d}\,\phi(\mathfrak{p},z)\ . \tag{19}$$

The compatibility of the two bounds on β imposes restrictions on the allowed range of $\mathfrak{p}$, namely, $\mathfrak{p}$ should satisfy the inequality

$$\mathfrak{p}^2 + \frac{\sqrt{d}\,\mathsf{Re}}{4\pi}\,\phi(\mathfrak{p},z) - 1 \ge 0\ . \tag{20}$$

In the interval $\mathfrak{p} \in [0,1]$, this inequality is satisfied for $\mathfrak{p} \in [p_{\min}(\mathsf{Re},z),1]$, where $p_{\min}(\mathsf{Re},z) \approx \frac{4\pi}{\sqrt{d}\,\mathsf{Re}}$ for large Re. Clearly, the lower bound on the range of $\mathfrak{p}$ is more meaningful for smaller Re.

2.4 Optimizing the Upper Bound on β

Since we do not have *a priori* control over $\mathfrak{p}$, we will derive an upper bound for β by maximizing the upper bound in (19) over $\mathfrak{p}$, after which we use the freedom in the choice of the parameter $z > 0$ to minimize for any given Re, which results in

$$\beta \le \min_{z>0}\ \max_{\mathfrak{p}\in[p_{\min}(\mathsf{Re},z),1]} \left[\frac{4\pi^2}{\mathsf{Re}}\,\mathfrak{p}^2 + 2\pi\sqrt{d}\,\phi(\mathfrak{p},z)\right]\ . \tag{21}$$

Although this procedure is not difficult to implement numerically, we will analyze only the case of high Re where the analysis can be carried out analytically. First notice that for high Re, the lower bound $p_{\min}(\mathsf{Re},z)$ is very small, so the maximization over $\mathfrak{p}$ can be taken in the entire interval $[0,1]$. Thus $\phi(\mathfrak{p},z) \le \phi^*(z) := \max_{\mathfrak{p}\in[0,1]} \phi(\mathfrak{p},z) = (1+z-z^2)^{-1/2}$ for $z \in [0, \frac{1+\sqrt{5}}{2})$. Since for large Re the Re-independent term in the right-hand side of (21) is dominating, we have the high-Re estimate

$$\beta \le \min_{z\in[0,\frac{1+\sqrt{5}}{2})} \left[\frac{4\pi^2}{\mathsf{Re}}\,\phi^*(z)^2 + 2\pi\sqrt{d}\,\phi^*(z)\right] = \frac{16\pi^2}{5\mathsf{Re}} + \frac{4\pi\sqrt{d}}{\sqrt{5}}\ . \tag{22}$$

At high Re, the value or $\mathfrak{p}$ maximizing $\phi(\mathfrak{p},z)$ is $\frac{2}{\sqrt{5}}$. We remark that it is not difficult to prove that the upper bound (22) is optimal (i.e., coincides with (21)) for $\mathsf{Re} \ge \frac{8\pi}{3\sqrt{5d}}$.

3 Discussion

In dimension 3, the scaling of the upper bound is in accord with conventional turbulence theory: at high Re, $\epsilon \sim \frac{U^3}{\ell}$ is independent of the molecular viscosity. For the type of forcing considered here, we find $\beta \le 4\pi\sqrt{\frac{3}{5}} \approx 9.7339\ldots$.

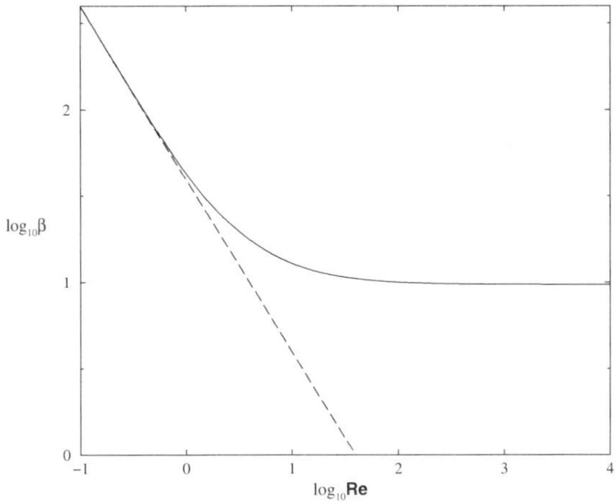

Fig. 1. Upper and lower bounds on β (solid and dashed lines, resp.)

A plot of the bounds is presented in Figure 1. At low Re, the upper and lower bounds converge to each other. While it is difficult to compare these bounds quantitatively with DNS results, we note from [7] that at high Re, values of β are typically around 1. Hence, our rigorous analysis, while yielding the expected scaling, overestimates the constants by about an order of magnitude.

In the 3-dimensional case, if we assume that the cascade is Kolmogorov, i.e., the spectral density of the energy is given by $E_{\mathrm{K}}(k) = C\epsilon^{2/3}k^{-5/3}$, we can estimate the "Kolmogorov" value $\mathfrak{p}_{\mathrm{K}}$ as follows:

$$E_{\text{kin, total}} \approx \int_{2\pi/\ell}^{\infty} E_{\mathrm{K}}(k)\,\mathrm{d}k\ , \quad E_{\text{kin, low}\,k} \approx \frac{2\pi}{\ell}\,E_{\mathrm{K}}\left(\frac{2\pi}{\ell}\right)\ ,$$

which gives $\mathfrak{p}_{\mathrm{K}} \approx \sqrt{\frac{2}{3}}$. Plugging this value in (21) and minimizing over z, we obtain the (approximate) estimate

$$\beta \le \frac{8\pi^2}{3\,\mathsf{Re}} + 2\sqrt{2}\pi \approx \frac{26.3}{\mathsf{Re}} + 8.9\ ,$$

which gives a slight improvement compared with the bounds (22).

Acknowledgments

CRD thanks B. Eckhardt, J. Schumacher, D. Lohse, K. Sreenivasan for stimulating and helpful discussions. This work was supported in part by NSF Award PHY-0244859.

References

1. J. Jiménez, A. A. Wray, P. G. Saffman, and R. S. Rogallo. The structure of intense vorticity in isotropic turbulence. *J. Fluid Mech.*, 255:65–90, 1993.
2. S. Grossmann and D. Lohse. Scale resolved intermittency in turbulence. *Phys. Fluids*, 6(2):611–617, 1994.
3. K. R. Sreenivasan. On the universality of the Kolmogorov constant. *Phys. Fluids*, 7(11):2778–2784, 1995.
4. L.-P. Wang, S. Chen, J. G. Brasseur, and J. C. Wyngaard. Examination of hypotheses in the Kolmogorov refined turbulence theory through high-resolution simulations. I. Velocity field. *J. Fluid Mech.*, 309:113–156, 1996.
5. P. K. Yeung and Y. Zhou. On the universality of the Kolmogorov constant in numerical simulations of turbulence. *Phys. Rev. E*, 56(2):1746–1752, 1997.
6. K. R. Sreenivasan and R. A. Antonia. The phenomenology of small-scale turbulence. In *Annual review of fluid mechanics, Vol. 29*, volume 29 of *Annu. Rev. Fluid Mech.*, pages 435–472. Annual Reviews, Palo Alto, CA, 1997.
7. K. R. Sreenivasan. An update on the energy dissipation rate in isotropic turbulence. *Phys. Fluids*, 10(2):528–529, 1998.
8. N. Cao, S. Chen, and G. D. Doolen. Statistics and structures of pressure in isotropic turbulence. *Phys. Fluids*, 11(8):2235–2250, 1999.
9. Y. Yamazaki, T. Ishihara, and Y. Kaneda. Effects of wavenumber truncation on high-resolution direct numerical simulation of turbulence. *J. Phys. Soc. Japan*, 71(3):777–781, 2002.
10. T. Gotoh, D. Fukayama, and T. Nakano. Velocity field statistics in homogeneous steady turbulence obtained using a high-resolution direct numerical simulation. *Phys. Fluids*, 14(3):1065–1081, 2002.
11. Y. Kaneda and T. Ishihara. Energy dissipation rate and energy spectrum in high resolution direct numerical simulations of turbulence in a periodic box. *Phys. Fluids*, 15(2):L21–L24, 2003.
12. J. Schumacher, K. R. Sreenivasan, and P. K. Yeung. Derivative moments in turbulent shear flows. *Phys. Fluids*, 15(1):84–90, 2003.
13. S. Childress, R. R. Kerswell, and A. D. Gilbert. Bounds on dissipation for Navier-Stokes flow with Kolmogorov forcing. *Phys. D*, 158(1-4):105–128, 2001.
14. C. R. Doering and C. Foias. Energy dissipation in body-forced turbulence. *J. Fluid Mech.*, 467:289–306, 2002.
15. C. R. Doering, B. Eckhardt, and J. Schumacher. Energy dissipation in body-forced plane shear flow. *J. Fluid Mech.*, 494:275–284, 2003.

New Potential Symmetries for a Generalised Inhomogeneus Nonlinear Diffusion Equation

M.L. Gandarias and S. Saez

Departmento de matematicas, Universidad de of Cadiz,
P.O. BOX 40, 11510, Puerto Real, Cadic, Spain
marialuz.gandarias@uca.es,
soledad.saez@uca.es

In this paper we consider a class of generalised diffusion equations which are of great interest in mathematical physics. For some of these equations that model fast diffusion new nonlocal potential symmetries are derived. These *nonclassical potential* symmetries allow us to increase the number of solutions. These solutions are not arising from classical potential symmetries.

1 Introduction

The diffusion processes appear in many physics processes such as plasma physics, kinetic theory of gases, solid state and transport in porous medium [1, 13]. One of the mathematical models for diffusion processes is the generalised inhomogeneous nonlinear diffusion equation

$$f(x)u_t = [g(x)(u^n u_x]_x. \tag{1}$$

In (1) $u(x,t)$ is a function of position x and time t and may represent the temperature, $f(x)$ and $g(x)$ are arbitrary smooth functions of position and may denote the density and the density-dependent part of thermal diffusion, respectively. The special case with $n = -1$ emerges in plasma physics and reveals a surprising richness of fast diffusion processes [13].

Lie classical symmetries admitted by nonlinear PDE's are useful for finding invariant solutions [11, 12]. Recently, the Lie classical symmetries for a variable coefficient nonlinear diffusion-convection equation have been derived in [5].

Bluman and Cole [3] developed the nonclassical method to study the symmetry reductions of the heat equation. In previous works, we have obtained nonclassical symmetries for a porous medium equation with absorption, and for a porous medium with convection [10].

In [4], Bluman introduced a method to find a new class of symmetries for a PDE when it can be written in a conserved form. These symmetries are nonlocal symmetries which are called *potential* symmetries. Potential symmetries were obtained in [7] for the porous medium equation when it can be written in a conserved form.

In a recent paper [14], C. Sophocleous has classified the potential symmetries of (1). He obtained that potential symmetries exists only if the parameter n takes the values -2 or $-\frac{2}{3}$ and also certain relations must be satisfied by the functions $f(x)$ and $g(x)$.

In [9], we have derived for the special case of (1), with $f(x) = 1$ and $g(x) = 1$ a new class of potential symmetries called *nonclassical potential* symmetries, which are realized as nonclassical symmetries of an associated system.

The aim of this work is to obtain for (1) *nonclassical potential* symmetries. The nonclassical method applied to the corresponding associated potential system lead to new symmetries *nonclassical potential* symmetries as well as to new exact solutions.

2 Nonclassical Potential Symmetries

It was pointed out in [14] that the transformation

$$x' = \int \frac{dx}{g(x)} = G(x),\ t' = t,\ u' = u \tag{2}$$

connects equation (1), and the pde

$$g(G^{-1}(x'))f(G^{-1}(x'))u_t = [(u^n u_{x'}]_{x'}, \tag{3}$$

where G^{-1} is the inverse function of G. Although in in [14] the author states that such choice does not simplify the analysis, it is clear that we can equivalently use an equation of the form (3) and then transform the results for equation (1) by using the corresponding point transformation.

Due to the fact that (2) connects equation (1) with (3) it happens that (2) also connects

$$\begin{aligned} v_x &= f(x)u \\ v_t &= g(x)u^{-1}u_x, \end{aligned} \tag{4}$$

with

$$\begin{aligned} v_{x'} &= h(x')u \\ v_t &= u^{-1}u_{x'}, \end{aligned} \tag{5}$$

We are now considering the nonclassical symmetries of (5).

We consider the associated auxiliary system given by (5) augmented with the invariance surface condition

$$\xi v_{x'} + \tau v_t - \psi = 0, \tag{6}$$

which is associated with the vector field

$$V = \xi(x', t, u, v)\partial_{x'} + \tau(x', t, u, v)\partial_t + \phi(x', t, u, v)\partial_u + \psi(x', t, u, v)\partial_v. \tag{7}$$

By requiring both (5) and (6) to be invariant under the transformation with infinitesimal generator (7) one obtains an over determined, nonlinear system of equations for the infinitesimals $\xi(x',t,u,v)$, $\tau(x',t,u,v)$, $\psi(x',t,u,v)$, $\phi(x',t,u,v)$. When at least one of the generators of the group depend explicitly of the potential , that is if

$$\xi_v^2 + \tau_v^2 + \phi_v^2 \neq 0 \tag{8}$$

then (7) yields a nonlocal symmetry of (1).
We can distinguish two different cases: $\tau = 0$ and $\tau \neq 0$.
If $\tau \neq 0$, we set $\tau = 1$ without loss of generality. The nonclassical method, with $\tau \neq 0$, applied to (5), give rise to nonlinear determining equations for the infinitesimals. If we require that $\xi_u = \psi_u = 0$, we obtain that

$$\phi = -\frac{\partial \xi}{\partial v}\, h\, u^2 + \left(\frac{\partial \psi}{\partial v} - \frac{\xi\, \frac{\partial h}{\partial x'}}{h} - \frac{\partial \xi}{\partial x'}\right) u + \frac{\frac{\partial \psi}{\partial x'}}{h} \tag{9}$$

and $h(x')$,$\xi(x',t,v)$ and $\psi(x',t,v)$ must satisfy the following conditions

$$\frac{\partial^2 \xi}{\partial v^2} - \xi\, \frac{\partial \xi}{\partial v} = 0$$

$$-h\,\frac{\partial^2 \psi}{\partial v^2} - \xi\, h\,\frac{\partial \psi}{\partial v} + \frac{\partial \xi}{\partial v}\, h\, \psi + \left(2\,\frac{\partial \xi}{\partial v}\,\frac{\partial h}{\partial x'} + 2\,\frac{\partial^2 \xi}{\partial v \partial x'}\, h\right) + \left(-\xi\,\frac{\partial \xi}{\partial x'} - \frac{\partial \xi}{\partial t}\right) h = 0$$

$$-\xi\, h^2\,\frac{\partial \psi}{\partial x'} - 2\, h^2\,\frac{\partial^2 \psi}{\partial v \partial x'} + h^2\,\psi\,\frac{\partial \psi}{\partial v} + h^2\,\frac{\partial \psi}{\partial t} + \frac{\partial \xi}{\partial x'}\, h^2 \psi$$

$$+\left(\xi\, h\,\frac{\partial^2 h}{\partial x'^2} - \xi\left(\frac{\partial h}{\partial x'}\right)^2 + \frac{\partial^2 \xi}{\partial x'^2}\, h^2\right) = 0$$

$$\left(h\,\psi + \frac{\partial h}{\partial x'}\right)\frac{\partial \psi}{\partial x'} - h\,\frac{\partial^2 \psi}{\partial x'^2} = 0$$

Despite the fact that the former equations are too complicated to be solved in general, special solutions can be obtained:
1. For

$$h(x') = k_3 e^{k_2 x'}, \quad \xi = -2\,\mathrm{k}_1\,\tanh(\mathrm{k}_1\, v + \mathrm{k}_4\, t + \mathrm{k}_5), \quad \psi = -2k_2 + \frac{k_4}{k_1}$$

2. For

$$h = k_2, \quad \xi = \mathrm{k}, \quad \psi = -2k_1 \tanh\,(\mathrm{k}_1(\mathrm{k}t + x')),$$

we obtain the similarity variable $z = x' - t$ and the family of invariant solutions is defined by

$$v = \frac{1}{2\mathrm{k}}\,(\log 2 - \log(\cosh(2\mathrm{k}_{\,1}(x'-t)) + 1) - \frac{1}{\mathrm{k}}\,(\log(\mathrm{sech}\,(\mathrm{k}_{\,1}(x'+t)\zeta_2)) + \mathrm{k}_3.$$

The corresponding exact solution of (1) adopts the form

$$u = -\frac{k_1^2}{k_2}\left(\tanh\left(k_1\int f(x)dx + k_2t\right) + \tanh\left(-k_1\int f(x)dx + k_2t\right)\right) \quad (10)$$

3. For

$$h(x') = k_3e^{k_2x'}, \quad \xi = k_1, \quad \psi = -2k_4\tanh\left(k_4(x' + k_1t + k_5)\right) - k_2,$$

we obtain that the similarity variable is $z = x' - k_1t$ and the family of invariant solutions for (1) with $f(x) = k_3e^{k_2x}$ and $g(x) = 1$ is defined by

$$u = -e^{-k_2x}\left(\tanh\left(\frac{2k_3x + k_3t + k_5}{4}\right) - \tanh\left(\frac{2k_3x - k_3t + k_5}{4}\right)\right) \quad (11)$$

We point out that although in cases [2,3], the infinitesimals do not depend on $v = \int u(x)dx$, and they do not correspond to a nonclassical potential symmetry, solutions (10), (11) are *new* exact solutions of (1) which *can not* be obtained by using classical potential symmetries of (1) [14].

These solutions, which describes an unusual diffusion process caused by flux suction at infinity, has been derived by P.Rosenau in [13] for $f(x) = 1$ and $g(x) = 1$ using a different procedure, and looks like the elastic interaction of two kinks giving a shrinking appearance to diffusion. Its derivative looks like the interaction of two solitons solutions. These are of special interest since such solutions are in general associated with integrable equations.

It is a pleasure to thank M. Oberlack for his useful suggestion.

References

1. Berryman J G and Holland C J (1982) J. Math. Phys. 23: 983.
2. Bluman G W (1992) Potential symmetries and linearization, in Proceedings of NATO Advanced Research Workshop, Kluwer, Exeter.
3. Bluman G W and Cole J (1969) J Math Mech 18: 1025.
4. Bluman G W and Kumei S (1980) J Math Phys 21: 1019.
5. El-labany S K, Elhanbaly A M and Sabry R (2002) J Phys A: Math and General 35: 8055–8063.
6. Gandarias M L (1996) J Phys A: Math and General 29: 607–633.
7. Gandarias M L (1996) J Phys A: Math and General 29: 5919–5934.
8. Gandarias M L (1997) J Phys A: Math and General 30: 6081.
9. Gandarias M L (2001) Phys Lett A, 286: 153–160.
10. Gandarias M L, Romero J L and Diaz J M (1999) J Phys A: Math and General 32: 1461.
11. Olver P J (1986) Applications of Lie Groups to Differential Equations (Berlin:Springer).
12. Ovsiannikov L V (1982) Group Analysis of Differential Equations New York: Academic Press.
13. Rosenau P (1995) Physical Review Letters 7: 1056–1059.
14. Sophocleous (2003) Physica A, 320: 169–183.

DNS and New Scaling Laws of ZPG Turbulent Boundary Layer Flow

G. Khujadze and M. Oberlack

Hydromechanics and Hydraulics Group, Darmstadt University of Technology, Petersenstr. 13, 64287 Darmstadt,
khujadze@hyhy.tu-darmstadt.de, oberlack@hyhy.tu-darmstadt.de

Lie group analysis is used to derive *new scaling laws* (exponential laws) for ZPG turbulent boundary layer flow. A new scaling group was found in the two-point correlation equations. DNS of such a flow was performed at $Re_\theta = 2240$ using a spectral method with up to 160 million grid points. The results of the numerical simulations are compared with the new scaling laws and good agreement is achieved.

1 Lie Group Analysis and New Scaling Laws

Lie group approach developed in [1, 2] applied to the fluctuation equations will presently be applied to the two-point correlation equations to find their symmetry groups. The present analysis is based on these equations with the restricted mean velocity profile $\bar{u}_1 \equiv \bar{u}_1(x_2)$, where x_2 is the wall normal coordinate. The governing equations will be considered in the outer part of boundary layer flow i.e. sufficiently apart from the viscous sublayer. For the present problem we focus only on scaling symmetries:

$$\mathbf{G_{s1}}: \quad \tilde{x}_2 = x_2 e^{c_1}, \ \tilde{r}_i = r_i e^{c_1}, \ \tilde{\bar{u}}_1 = \bar{u}_1 e^{c_1}, \ \tilde{R}_{ij} = R_{ij} e^{2c_1}, \cdots \tag{1}$$

$$\mathbf{G_{s2}}: \quad \tilde{x}_2 = x_2, \ \tilde{r}_i = r_i, \ \tilde{\bar{u}}_1 = \bar{u}_1 e^{-c_2}, \ \tilde{R}_{ij} = R_{ij} e^{-2c_2}, \cdots \tag{2}$$

$$\mathbf{G_{s3}}: \quad \tilde{x}_2 = x_2, \ \tilde{r}_i = r_i, \ \tilde{\bar{u}}_1 = \bar{u}_1, \ \tilde{R}_{ij} = R_{ij} e^{-c_3}, \cdots \tag{3}$$

The variables $c_1 - c_3$ are the group parameters of the corresponding transformations. The dots denote that also higher order correlations are involved in the corresponding transformations.

The most interesting fact with respect to the latter groups is that *three independent scaling groups* $\mathbf{G_{s1}}$, $\mathbf{G_{s2}}$, $\mathbf{G_{s3}}$ have been computed. Two symmetry groups, namely (1) and (2), correspond to the scaling symmetries of the Euler equations. The first one is the scaling in space, the second one scaling in time. The third group ($\mathbf{G_{s3}}$) is a *new scaling group* that is a characteristic feature only of the one-dimensional turbulent boundary layer flow. This is in striking contrast to the Euler and Navier-Stokes equations, which only admit two and one scaling groups, respectively.

The corresponding characteristic equations for the invariant solutions (see [1]) read

$$\frac{dx_2}{c_1x_2+c_4}=\frac{dr_i}{c_1r_i}=\frac{d\bar{u}_1}{(c_1-c_2)\bar{u}_1+c_5}=\frac{dR_{ij}}{[2(c_1-c_2)+c_3]R_{ij}}=\cdots, \tag{4}$$

where c_4 and c_5 correspond to translation in space and Galilean invariance respectively.

Imposing the assumption of symmetry breaking of the scaling of space ($c_1 = 0$) (for details see [1]) an extended set of scaling laws for the mean velocity and two-pont correlations were obtained as follows:

$$\bar{u}_1(x_2)=k_1+k_2e^{-k_3x_2}, \qquad R_{ij}(x_2,\boldsymbol{r})=e^{-k_4x_2}B_{ij}(\mathbf{r}), \tag{5}$$

where $k_1 \equiv c_5/c_2$, $k_3 \equiv c_2/c_4$ and k_2 is a constant of integration. $k_4 \equiv 2c_2 - c_3/c_14$ is a constant comprising several group parameters and constants of integration and B_{ij} is a function of $\boldsymbol{r}$ only.

In normalized and nondimensional variables the exponential scaling law (5) may be re-written in the following form

$$\frac{\bar{u}_\infty-\bar{u}}{u_\tau}=\alpha\exp\left(-\beta\frac{x_2}{\Delta}\right), \qquad \frac{\overline{u_i'u_j'}(x_2)}{u_\tau^2}=b_{ij}\exp\left(-a\frac{x_2}{\Delta}\right), \tag{6}$$

where α and β are dimensionless parameters and $\Delta = \frac{\bar{u}_\infty}{u_\tau}\delta^*$ is the Rotta-Clauser length scale, while δ^* is the boundary layer displacement thickness. The scaling laws for the Reynolds stresses were derived from equation (5) at $\mathbf{r} = 0$. b_{ij} and a are universal constants that should be found from DNS or experiments.

2 DNS of Turbulent Boundary Layer Flow

The code for the DNS was developed at KTH, Stockholm (for details see [3]) using a spectral method. The simulation was done with $N \approx 158$ million grid points. All quantities are non-dimensionalised by the free-stream velocity u_∞ and δ^* at $x = 0$. The size of the computational box is $450\delta^* \times 30\delta^* \times 34\delta^*$. Simulation was run for $4000\delta^*/u_\infty$ time units. The statistics accumulation was performed during the last 3000 time units. The useful region was confined to $150 - 300\delta^*|_{x=0}$ which corresponds to Re_θ from 1670 to 2240. Reynolds stress components are shown in Fig. 1 (left plot). Results obtained by [4] are compared to the present DNS data. For $\overline{u_1'u_1'}$ the peak is higher for present DNS: $(\overline{u_1'u_1'})_{peak} \approx 7.48$ at $y^+ \approx 13.87$.

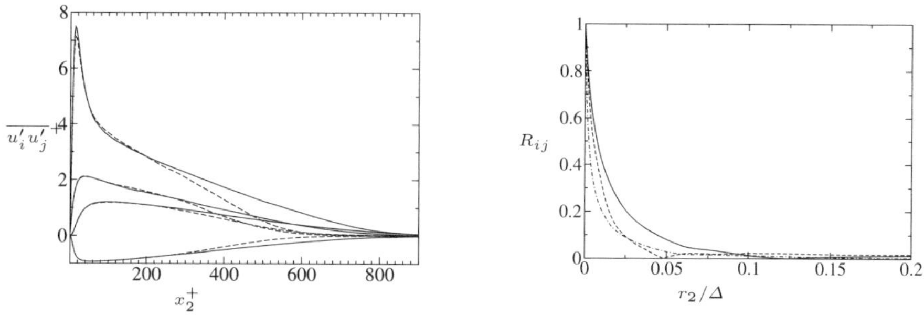

Fig. 1. *Left figure:* $\overline{u'_i u'_j}$ in plus units for $Re_\theta = 1410, 2240$, from Spalart's (- - -) and present DNS (——). *Right figure:*—— R_{12}, - - - R_{21} and —·— R_{22} for $Re_\theta = 2240$.

In addition, two-point correlation functions were calculated and $R_{12}(x_2, r_2)$, $R_{21}(x_2, r_2)$ and $R_{22}(x_2, r_2)$ are plotted against $r_2 = x'_2 - x_2$ normalized by the corresponding Reynolds stresses:

$$R_{ij}(x_2, r_2) = \frac{\overline{u'_i(x_2)u'_j(x_2 + r_2)}}{\overline{u'_i(x_2)u'_j(x_2)}}$$

Two-point correlation functions R_{12}, R_{21} and R_{22} are shown in Fig. 1 (right plot) for $Re_\theta = 2240$ at point $x_2/\Delta = 0.56$.

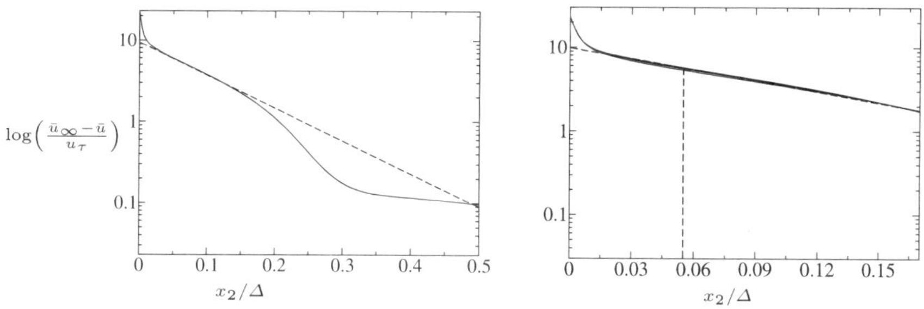

Fig. 2. Mean velocity profile in log-linear scaling (*left plot*). Close up (*right plot*) view at different Re_θ. - - - theoretical (Eq. 6) result and —— present DNS.

The mean velocity of the turbulent boundary layer data is plotted in Fig. 2. As it is observed from the figure, DNS and theoretical results (Eq. 6) are in good agreement in the region $x_2/\Delta \approx 0.01 - 0.15$. After this region the velocity defect law decreases more rapidly than what was derived from the theoretical result. There may be two reasons for this behavior. First it might be the result of the low Reynolds number phenomenon in the DNS while the theoretical results were obtained for the large Reynolds number limit. Second, the non-parallel effects become dominant in the outer part of the wake region, that

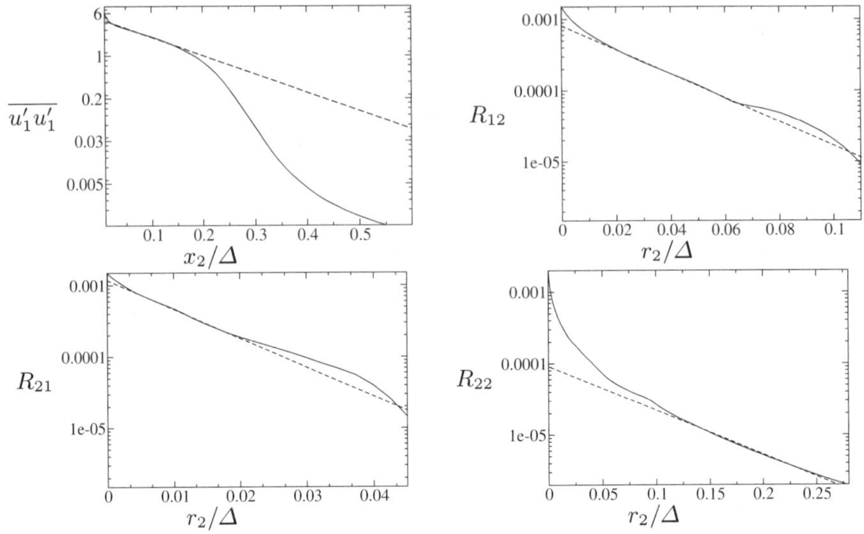

Fig. 3. $\overline{u_1'u_1'}$ and R_{12}, R_{12} and R_{22} versus x_2/Δ. —— DNS, - - - theoretical result.

could lead to the deviation from the exponential law, because the theoretical results were derived using fully parallel flow assumption (see [5]). The close up plot of mean velocity profiles is presented in Fig. 2 (right plot) at different Reynolds numbers $Re_\theta = 1670, 1870, 2060, 2240$. Good collapse of profiles in the exponential region is seen from the plot.

In Fig. 3 $\overline{u_1'u_1'}$ and R_{12}, R_{21}, R_{22} are presented. The theoretical results are compared to the DNS. Constants in the exponential scaling laws (see Eqs. (5) and (6)) are different for the different variables. The last means that the "coincidence region" for each variable is located in the different area of the outer part of the boundary layer flow.

Conclusion: New symmetries of the two-point correlation equations are found which were used to derive new scaling laws for two-point and Reynolds stress quantities. DNS of turbulent boundary layer flow was performed for $Re_\theta = 2240$ which shows the validity of the obtained new scaling laws.

References

1. Oberlack, M. (2001). *J.Fluid Mech.*, **427**, pp. 299-328.
2. Oberlack, M. (2000). *Z.Angew.Math.Mech*, **80** 11-12, 791–800.
3. Lundbladh *et al.* (1999). *Tech. Rep.* 1999:11. KTH, Stokholm.
4. Spalart, P.R. (1988). *J. Fluid Mech.* **187**, 61–98.
5. Lindgren B., (2002). *Doctoral thesis*, KTH, Stockholm.

Linear Instability of a Slowly Divergent Planar Jet

Polina Landa[1], Vadim Ushakov[1,2], and Jürgen Kurths[2]

[1] Department of Physics, Lomonosov Moscow State University, 119899 Moscow, Russia planda@lpi.ru uvadim@mail.ru
[2] Nonlinear Dynamics Group, Department of Physics Potsdam University, D-14415 Potsdam, Germany juergen@agnld.uni-potsdam.de

Summary. By using small deviations from the stationary solution of the Navier–Stokes equation, the problem of linear instability of a plane submerged subsonic jet is considered. In the approximation of weak divergence of the jet, this problem reduces to a linear not self-adjoint boundary value problem with a given behavior of the variables at large values of the transversal coordinate. The solution of this boundary value problem allow us to calculate the gain factor and the phase velocity of hydrodynamical waves as functions of frequency and of distance from the nozzle. We have found that the dependence of the gain factor on the frequency has a resonant character. As the distance from the nozzle increases, the dependence of the gain factor on the frequency becomes more narrow and the maximum of that shifts into the small frequency region. Hence, the hydrodynamical waves become more and more coherent. The obtained results are in good agreement with experimental data.

1 Introduction

We consider a plane jet issuing from a nozzle of width $2d$. Neglecting compressibility, we may describe the processes in such a jet by the two-dimensional Navier–Stokes equations for the stream function $\Psi(t,x,y)$ and vorticity $\Omega(t,x,y)$. In the dimensionless coordinates these equations are

$$\begin{aligned}
&\Omega(t,x,y) = \Delta\Psi(t,x,y)\,,\\
&\frac{\partial\Omega(t,x,y)}{\partial t} - \frac{\partial\Psi(t,x,y)}{\partial x}\,\frac{\partial\Omega(t,x,y)}{\partial y}\\
&+\frac{\partial\Psi(t,x,y)}{\partial y}\,\frac{\partial\Omega(t,x,y)}{\partial x} - \frac{2}{\mathsf{Re}}\,\Delta\Omega(t,x,y) = 0\,,
\end{aligned} \tag{1}$$

where Δ is the Laplacian, $\mathsf{Re} = 2U_0 d/\nu$ is the Reynolds number and U_0 is the longitudinal velocity component in the center of the nozzle. The stream function $\Psi(t,x,y)$ is related to the longitudinal (U) and transversal (V) components of the flow velocity by $U(t,x,y) = \partial\Psi/\partial y, \quad V(t,x,y) = -\,\partial\Psi/\partial x$.

This way we also get a dimensionless time and the circular frequencies are measured in $\mathsf{S} = \omega d/U_0 \equiv \pi\mathsf{St}$, where St is the Strouhal number.

In accordance with the idea on the nature of turbulence in nonclosed fluid flows [1, 2], for the onset of turbulence some source of noise is necessary. This source really exists at the nozzle exit. We can split a solution of (1) into dynamical and stochastic constituents and set:

$$U(t,x,y) = u_{\mathrm{d}}(x,y) + \frac{\partial\psi(t,x,y)}{\partial y}\,, \quad V(t,x,y) = v_{\mathrm{d}}(x,y) - \frac{\partial\psi(t,x,y)}{\partial x}\,,$$
$$\Omega(t,x,y) = \Omega_{\mathrm{d}}(x,y) + \Delta\psi(t,x,y)\,, \tag{2}$$

where $u_{\mathrm{d}}(x,y)$, $v_{\mathrm{d}}(x,y)$ and $\Omega_{\mathrm{d}}(t,x,y)$ are dynamical constituents of velocity and vorticity, which are the solutions of the stationary Navier–Stokes equations.

In the case of a laminar issue from the nozzle, we set the profile of dynamical longitudinal velocity such that, at the nozzle exit, the shape of the mixing layer is close to that which is described by the Blasius equation [3]. Taking into account the entrainment of some ambient fluid, we set:

$$u_{\mathrm{d}}(x,y) = \frac{1}{2}\left[1 - \tanh\left(q\,\frac{|y|-1}{\delta_0(x)} - r(x)\right)\right],$$

where $q = 3$, $\delta_0(x)$ and $r(x)$ are unknown functions of x, $\delta_0(x) = \delta_1(x) + \delta_2(x)$ is the mixing layer thickness. The internal and external boundaries of the mixing layer are conditionally found from the relations: $u_0(x, 1-\delta_1) = \alpha$ and $u_0(x, 1+\delta_2) = 1-\alpha$, where α is a certain number close to 1. The dynamical

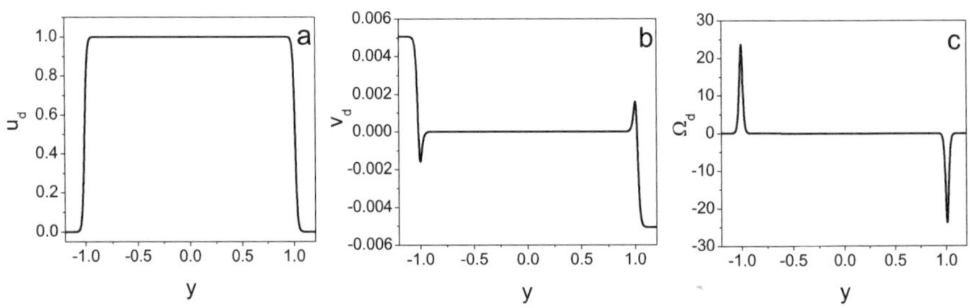

Fig. 1. Profiles of $u_{\mathrm{d}}(0,y)$ (**a**), $v_{\mathrm{d}}(0,y)$ (**b**) and $\Omega_{\mathrm{d}}(0,y)$ (**c**).

constituents of the transversal velocity and vorticity can be obtained from the continuity equation and the equation for $\Omega_{\mathrm{d}}(x,y)$ respectively (see fig. 1).

To find $\delta_0(x)$ and $r(x)$, we used the conservation laws for the fluxes of momentum and energy [4]. As a result we obtain, that within the initial part of the jet, $r(x) \approx r_0 = 0.5$ and depends on $\delta_0(x)$ only slightly, $\delta_1(x) \approx \delta_0(x)/3$, $\delta_2(x) \approx 2\delta_0(x)/3$, $\delta_0(x) \approx \sqrt{(32\,q^2 x)/(3\mathsf{Re}) + \delta_0^2(0)}$.

2 Linearized Equations

Substituting (2) in (1) and linearizing (1) for stochastic constituents we get:

$$\begin{aligned} &\frac{\partial \Delta\psi}{\partial t} + u_{\mathrm{d}}(x,y)\,\frac{\partial \Delta\psi}{\partial x} + v_{\mathrm{d}}(x,y)\,\frac{\partial \Delta\psi}{\partial y} \\ &-\Omega_{\mathrm{d}y}(x,y)\,\frac{\partial \psi}{\partial x} + \Omega_{\mathrm{d}x}(x,y)\,\frac{\partial \psi}{\partial y} - \frac{2}{\mathsf{Re}}\,\Delta\Delta\psi = 0\,. \end{aligned} \tag{3}$$

We suppose that the jet diverges slowly and seek a partial solution of (3) in the form of running hydrodynamical waves of the frequency S with a slowly varying complex wave number:

$$\begin{aligned} \psi(t,x,y) = &\left(f_0(\mathsf{S},\tilde{x},y) + \frac{1}{\lambda}\,f_1(\mathsf{S},\tilde{x},y) + \dots\right) \\ &\times \exp\left[i\mathsf{S}t - i\int\limits_0^x \left(Q_0(\mathsf{S},\tilde{x}) + \frac{Q_1(\mathsf{S},\tilde{x})}{\lambda} + \dots\right) d\tilde{x}\right], \end{aligned} \tag{4}$$

where S is the wave frequency in the time scale given by us, $Q(\mathsf{S},\tilde{x}) = Q_0(\mathsf{S},\tilde{x}) + Q_1(\mathsf{S},\tilde{x})/\lambda + \dots$ is the complex wave number, $\tilde{x} = x/\lambda$ is the "slow" coordinate, $\lambda = \sqrt{\mathsf{Re}}$, $f(\mathsf{S},\tilde{x},y)$ is an unknown function vanishing, along with its derivatives with respect to y, at $y = \pm\infty$. Equating the terms of the order of $1/\lambda$ and retaining only terms with the first derivatives with respect to $\tilde{x}$, we yield the following equations for $f_0(\mathsf{S},\tilde{x},y)$ and $f_1(\mathsf{S},\tilde{x},y)$:

$$\mathcal{L}_0(Q_0)f_0 = 0, \tag{5}$$

$$\mathcal{L}_0(Q_0)f_1 = Q_1\mathcal{L}_1(Q_0)f_0 - \mathcal{L}_2(Q_0)f_0\,, \tag{6}$$

where $\mathcal{L}_0(Q_0)$, $\mathcal{L}_1(Q_0)$ and $\mathcal{L}_2(Q_0)$ are certain not self-adjoint differential operators. We restrict ourself to the first approximation (i.e. we use only (5)).

Equation (5), with the boundary conditions indicated above, describes a non self-adjoint boundary-value problem, where Q_0 plays the role of an eigenvalue. It is convenient to solve (5) separately for three regions: Over region 1 ($|y| \le y_1(x)$), where $u_{\mathrm{d}}(x,y) \approx 1$, $v_{\mathrm{d}}(x,y) \approx 0$ and $\Omega_{\mathrm{d}}(x,y) \approx 0$, and over region 2 ($|y| \ge y_2(x)$), where $u_{\mathrm{d}}(x,y) \approx 0$, $v_{\mathrm{d}}(x,y) = v_{\mathrm{d}}(x,\pm\infty)$, $\Omega_{\mathrm{d}x}(x,y) \approx \Omega_{\mathrm{d}x}(x,\pm\infty)$ and $\Omega_{\mathrm{d}y}(x,\pm\infty) \approx 0$), we have found general analytical solutions of (5) with arbitrary constants. Here $y_1(x)$ and $y_2(x)$ are internal and external boundaries of the mixing layer. Over region 3 ($y_1(x) \le |y| \le y_2(x)$) all terms of (5) are of the same order. In this region the general solution can be written as a sum of fundamental solutions, which are obtained numerically, multiplying by an arbitrary constants. Sewing this solution and its derivatives with general solutions, obtained in regions 1 and 2, at the points $y = y_1(x)$ and $y = y_2(x)$ correspondingly, we get a system of linear equations for the arbitrary constants.

The eigenvalues can be found from the requirement that the determinant of the system of linear equations is equal to zero. This determinant contains the results of the subtraction of a large numbers. As a consequence, the value of the determinant is random. To avoid this problem, we solve the differential equations for these differences, instead of the initial equations. As a result, we find the eigenvalues $Q_0(\mathsf{S}, \tilde{x})$, which are represented in Fig. 2. We set $Q_0(\mathsf{S}, \tilde{x}) = \mathsf{S}/v_{\rm ph}(\mathsf{S}, \tilde{x}) + \imath G(\mathsf{S}, \tilde{x})$, where $v_{\rm ph}(\mathsf{S}, \tilde{x})$ is the phase velocity of a propagating hydrodynamical wave, and $G(\mathsf{S}, \tilde{x})$ is the gain factor of this wave. For each S we take the eigenvalue, corresponding to the maximal gain factor.

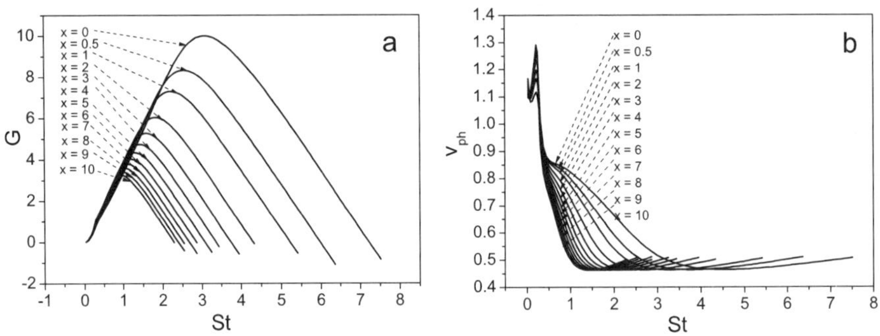

Fig. 2. Dependence of $G(\mathsf{S}, \tilde{x})$ and $v_{\rm ph}(\mathsf{S}, \tilde{x})$ on the Strouhal number, for a different distances from the nozzle.

It is seen from Fig. 2 that the gain factor maximum is shifted to the low-frequency region as the distance from the nozzle increases. This effect corresponds to the phenomenon of a shift of the velocity power spectrum, which is observed experimentally. Usually this phenomenon is explained by the vortex paring [5], however it can be obtained in the framework of our linear theory. In line with this theory, the experimentally observed shift of the power spectrum is caused by the divergence of the jet, but not by paring of vortices.

References

1. Landa P S (1996) Europhys. Lett. 36:401
2. Landa P S, Zaikin A A, Ginevsky A S and Vlasov Ye V (1999) Int. J. Bifurcation and Chaos 9:397
3. Landau L D, Lifshitz E M (1986) Hydrodinamics. Naika, Moscow (in Russian)
4. Landa P S (1996) Nonlinear Oscillations and Waves in Dynamical Systems. Kluwer Academic Publ., Dordrecht
5. Belotserkovsky S M, and Ginevsky A S (1995) Simulation of Turbulent Jets and Wakes by Dscrete Vortex Technique Nauka, Moscow (in Russian)

On How Different Are Genuine and 'Passive' Turbulence

Arkady Tsinober

Faculty of Engineering, Tel Aviv University, 69978 Tel Aviv, Israel
tsinober@eng.tau.ac.il

One of the common views is that the behavior of passive objects in turbulent flows is in many respects similar to that of genuine turbulence. However, there exist a number of essential qualitative differences which require caution in promoting analogies between the two and which may be misleading. Some of these differences were summarized in [1]. It is the purpose of this note to provide a critical update of these essential differences with the emphasis on the aspects not given in [1] including the new results appeared after 2001.

1 Examples of Proposed Analogies

The story starts with the Reynolds analogy in 1874 on transport of momentum and heat[2] and his proposal to study fluid motion by means of 'color bands' [3], which can be seen as the foundation of flow visualization. The next example concerns the frozenness of vorticity in the flow field in inviscid flows thereby indicating the analogy between vorticity and (infinitesimal) material lines [4], [5]. Batchelor [6] proposed an analogy between amplification of vorticity and magnetic field in a turbulent flow of conducting fluid. Recent statements are made in the same spirit. For example, it is that *the advection-diffusion equation, in conjunction with a velocity field model with turbulent characteristics ,serves as a simplified prototype problem for developing theories for turbulence itself* [7], and that *the well established phenomenological parallels between the statistical description of mixing and fluid turbulence itself suggest that progress on the latter front may follow from a better understanding of turbulent mixing* Shraiman and Siggia [8].

2 Main Differences

However, *momentum is not a passive contaminant; "mixing" of mean momentum relates to the dynamics of turbulence, not merely its kinematics* [9].

In other words, the differences are more than essential: the evolution of passive objects is not related to the dynamics of turbulence in the sense that the dynamics of fluid motion does not enter in the problems in question - the velocity field is prescribed *a priory* in all problems on evolution of passive objects. Consequently the problems associated with the passive objects are *linear*; whereas genuine turbulence is a strongly nonlinear problem - nonlinearity is in the heart of turbulent flows and is underlying the main manifestations of the differences between genuine and passive turbulence.

2.1 Is There a 'Cascade' of Passive Objects?

It is rather common to speak about cascade in case of passive objects. The main argument is from some analogy, e.g. [8]. Indeed, for instance in any random isotropic flow the rate of production of 'dissipation' (i.e. corresponding field of derivatives) of both passive scalars and passive vectors is essentially positive, which can be interpreted as a sort of 'cascade'. However, the equations describing the behavior of passive objects are linear. Hence, there is no interaction between modes of whatever decomposition of the field of a passive object: the principle of superposition is valid in case of passive objects[1]. Therefore, more appropriate is to describe the process in terms of production of the field of derivatives of the passive object, which is performed by the velocity straining field, just like in case of velocity field [1]. Hence the extension of Kolmogorov arguments and phenomenology to passive objects seems to be not justified. No wonder that the phenomenological paradigms for the velocity field failed in most cases when applied to passive objects.

2.2 Self-amplification of Velocity Derivatives

Nonlinearity of genuine turbulence is the reason for the self-amplification of the field of velocity derivatives, both vorticity and strain. In contrast there is no phenomenon of self-amplification in the evolution of passive objects[2]. It should be stressed that the process of self-amplification of strain is a specific feature of the dynamics of genuine turbulence having no counterpart in the behavior of passive objects. In contrast, the process of self-amplification of

[1] Here 'mode' is meant as a *solution* of the appropriate (linear) equation, e.g. of the advection-diffusion equation. Use 'modes' that are *not* solutions of this equation, such as Fourier modes, results in their interaction, since one of the coefficients of the advection-diffusion equation, the velocity field, is not constant. This interaction is interpreted frequently as a 'cascade' of passive objects. Such interaction is decomposition dependent, and therefore is not appropriate for description of physical processes, which are invariant of our decompositions.

[2] Such as material lines, gradients of passive scalar and solenoidal passive vectors with finite diffusivity

vorticity, along with essential differences[3], has common features with analogous processes in passive vectors; in both the main factor is their interaction with strain, whereas the production of strain is much more 'self' [1]. A related important difference is absence of pressure in case of passive objects.

2.3 Difference in Structure(s)

Along with some common features the mechanisms of formation of structure(s) are essentially different for the passive objects and the dynamical variables. One of the reasons is the presence of Lagrangian chaos, which is manifested as rather complicated structure of passive objects even in very simple regular velocity fields[4]. Therefore one can expect that the structure of passive objects in turbulent flows arises from two (essentially inseparable) contributions: one due to the Lagrangian chaos and the other due to the random nature of the velocity field itself [5]. Among other reasons are differences in sensitivity to initial (upstream) conditions (i.e. Lagrangian 'memory'), 'symmetries', e.g. the velocity field may be locally isotropic, whereas the passive scalar may not be and some other (see references in [1]. A recent result [10] for turbulent flow in a plane channel is an interesting addition to the list of these differences: although the vortical structure of the flow is the same, the scalar streak spacing varies by an order of magnitude depending on the mean profile of the scalar concentration. Moreover, passive scalar streaks were observed even in an artificial "structureless" flow field.

2.4 Kolmogorov 4/5 versus Yaglom 4/3 Laws

The Kolmogorov and the Yaglom laws are respectively $S_3(r) \equiv \left\langle (\Delta u_{\|})^3 \right\rangle = -\frac{4}{5}\epsilon r$, and $\left\langle \Delta u_{\|}(\Delta\theta)^2 \right\rangle = -\frac{4}{3}\epsilon_\theta r$, where $\Delta u_{\|} \equiv [\mathbf{u}(\mathbf{x}+\mathbf{r}) - \mathbf{u}(5)] \cdot \mathbf{r}/r$, $\Delta\theta = \theta(\mathbf{x}+\mathbf{r}) - \theta(\mathbf{x})$, ϵ - is the rate of dissipation of kinetic energy and $\epsilon_\theta = D\frac{\partial\theta}{\partial x_i}\frac{\partial\theta}{\partial x_i}$ - is the rate of dissipation of fluctuations of a passive scalar.

[3] We would like to stress again that vorticity is an active vector, since it 'reacts back' on the velocity (and thereby on strain) field. This is not the case with passive objects - the process here is 'one way': the velocity field does not 'know' anything about the passive object.

[4] On the other hand, e.g. the ramp-cliff structures of a passive scalar are observed in pure Gaussian 'structureless' random velocity field, just like those in a variety of real turbulent flows practically independently of the value of the Reynolds number.

[5] Therefore one cannot claim that *statistical properties of this so-called 'passive scalar' turbulence are decoupled from those of the underlying velocity field* [8], *since the non-trivial statistical properties of scalar turn out to originate* not only *in the mixing process itself* but are *inherited from the complexity of the turbulent velocity field* as well. *Study of passive scalar turbulence is therefore* not *decoupled from the still intractable problem of calculating the velocity statistics...*

The analogy between these two laws[6], though useful in some respects [11] is violated for a Gaussian velocity field. Namely, the 4/3 law remains valid for such (as any other random isotropic) velocity field, whereas the 4/5 law is not, because $S_3(r) \equiv 0$ for a Gaussian velocity field. This difference is one of the manifestations of the dynamical nature of the Kolmogorov law as contrasted to the kinematical nature of the Yaglom law. It reflects the difference between genuine turbulence as a dynamical phenomenon and the 'passive' turbulence as a kinematical process.

2.5 Vorticity versus Passive Vectors. Material Lines

The common analogy is between vorticity in an inviscid flow and material lines. It is true that in the absence of viscosity vortex lines are material lines. However, these material lines are special in several respects. First, they are not passive as all the other material lines which are passive. In other words, the fact that vorticity is frozen in the inviscid flow field does not mean that vorticity is the same as material lines, but the other way around: those material lines which coincide with vorticity are like vorticity, because they are not passive anymore as other material lines. Second, these special material lines form a solenoidal vector field, whereas other material lines comprise infinitely many vector fields which are not necessarily divergent-free and which conserve the $div\mathbf{l}$ for infinitesimally small $\mathbf{l}$'s. For more details see [1] and [12].

2.6 Vorticity versus Passive Vectors. Solenoidal Vector Fields with Nonvanishing Diffusivity

The usual comparison is based on looking at the equations for vorticity ω and the (solenoidal) passive vector, $\mathbf{B}$, e.g. magnetic field in electrically conducting fluids [6]

$$\frac{\partial \omega}{\partial t} = \nabla\times(\mathbf{u} \times \omega) + \nu\nabla^2\omega, \quad \frac{\partial \mathbf{B}}{\partial t} = \nabla\times(\mathbf{u} \times \mathbf{B}) + \eta\nabla^2\mathbf{B} \qquad (1a, 1b)$$

Though a number of differences are known [1] these differences are hidden when one looks at the equations for ω and $\mathbf{B}$, which look quite 'similar'. However, a more 'fair' comparison should be made between the velocity field, $\mathbf{u}$, and the vector potential $\mathbf{A}$, with $\mathbf{B} = \nabla \times \mathbf{A}$, [13]. Such a comparison allows to see immediately one of the basic differences between the fields $\mathbf{u}$ and $\mathbf{A}$ (apart of the first obeying nonlinear and the second linear equation) which is not seen from the equations (1). Namely, the Euler equations conserve energy, since the scalar product of $\mathbf{u}\cdot(\omega \times \mathbf{u}) \equiv \mathbf{0}$. In contrast - unless initially and thereby subsequently $\mathbf{u} \equiv \mathbf{A}$ – the scalar product of $\mathbf{A} \cdot (\mathbf{u} \times \mathbf{B}) \not\equiv \mathbf{0}$ (the coresponding equation for the vector potential $\mathbf{A}$ has the form $\partial \mathbf{A}/\partial t +$

[6] The 4/5 Kolmogorov law follows by isotropy from the the 4/3 law for the velocity field in the form $\langle \Delta u_{\|}(\Delta \mathbf{u})^2 \rangle = -\frac{4}{3}\langle \epsilon \rangle r$.

$\mathbf{B} \times \mathbf{u} = -\nabla p_A$). It is this term $\mathbf{A} \cdot (\mathbf{u} \times \mathbf{B}) \equiv -A_i A_k s_{ik} + \partial/\partial x_k \{A_k A_l u_l - \frac{1}{2} u_k A^2\}$ which acts as a production term in the energy equation for $\mathbf{A}$. In other words the field $\mathbf{A}$ (and $\mathbf{B}$), is sustained by the strain, s_{ik}, of the velocity field - in contrast to the field $\mathbf{u}$. This leads, in particular, to substantial differences in amplification of vorticity, ω and $\mathbf{B}$, e.g. in statistically stationary velocity field (both NSE and Gaussian) the enstrophy ω^2 saturates to some constant value, whereas the energy of magnetic field B^2 grows without limit. For more details and other results concerning differences between ω and $\mathbf{B}$ see [13].

2.7 Vorticity versus Passive Vectors. Evolution of Disturbances

Important aspects of the essential difference between the evolution of fields ω and $\mathbf{B}$ arising from the nonlinearity of the equation of ω and linearity of the equation for $\mathbf{B}$ are revealed when one looks at how these fields amplify disturbances. The reason is that the equation for the disturbance of vorticity differ strongly from that for vorticity itself due to the nonlinearity of the equation for the undisturbed vorticity ω, whereas the equation for the evolution of the disturbance of $\mathbf{B}$ is the same as that for $\mathbf{B}$ itself due to the linearity of the equation for $\mathbf{B}$. Consequently, the evolution of disturbances of the fields ω and $\mathbf{B}$ is drastically different. For example, in a statistically stationary velocity field the energy of the disturbance of $\mathbf{B}$ grows exponentially without limit (just like the energy of $\mathbf{B}$ itself), whereas the energy of vorticity disturbance grows much faster than that of $\mathbf{B}$ for some initial period until it saturates at a value which is of order of the enstrophy of the undisturbed flow. It is noteworthy that much faster growth of the energy of disturbances of vorticity during the very initial (linear in the disturbance) regime is due to additional terms in the equation for the disturbance of vorticity, which have no analogues in the case of passive vector $\mathbf{B}$. It is important to stress that these additional 'linear' terms arise due to the nonlinearity of the equations for the undisturbed vorticity. In this sense the essential differences between the evolution of the disturbances of vorticity and the evolution of the disturbance of passive vector $\mathbf{B}$ with the same diffusivity can be seen as originating due to the nonlinear effects in genuine NSE turbulence even during the linear regime. For more details and other results concerning differences between the evolution of disturbances of ω and $\mathbf{B}$ see [13].

3 Concluding Remarks

The essential differences in the behavior of passive and active fields described above point to serious limitations on analogies between the passive and active fields and show that caution is necessary in promoting such analogies. They also serve as a warning that flow visualizations used for studying the structure of dynamical fields (velocity, vorticity, etc.) of turbulent flows may be quite misleading, making the question "what do we see?" extremely nontrivial. The general reason is that the passive objects may not 'want' to follow the dynamical fields (velocity, vorticity, etc.) due to the intricacy of the relation between passive and active fields just like there is no one to one relation between the Lagrangian and Eulerian statistical properties in turbulent flows [1]. This does not mean that qualitative and even quantitative study of fluid motion by means of 'color bands' [3] is impossible or necessarily erroneous. However, watching the dynamics of material 'colored bands' in a flow may not reveal the nature of the underlying motion, and even in the case of right qualitative observations the right result may come not necessarily for the right reasons. The famous verse by Richardson belongs to this kind of observation. On the other hand there are properties of passive objects which do depend on the details of the velocity field [1], [13]. Just these very properties can be effectively used to study the differences between the real turbulent flows and the artificial random fields, to gain more insight into the dynamics of real turbulence. At present, however, the knowledge necessary for such a use is very far from being sufficient. With few exceptions it is even not clear what can be learnt about the dynamics of turbulence from studies of passive objects (scalars and vectors) in real and 'synthetic' turbulence. This requires systematic comparative studies of both. As an attempt of such a comparative study was made by Tsinober and Galanti [13]. This is a relatively small part of a much broader field of comparative study of 'passive' turbulence reflecting the kinematical aspects and genuine turbulence representing also the dynamical processes.

References

1. Tsinober A. (2001) An informal introduction to turbulence, Kluwer, Dordrecht.
2. Reynolds, O. (1874) Proc. Lit. Phil. Soc. Manchester, 14: 7-12.
3. Reynolds, O. (1894) Nature, 50: 161-164.
4. Helmholz, H.(1858), Engl.transl. (1867) London Edinburgh Dublin Phil. Mag. J. Sci., Fourth series, 33: 485-512.
5. Kelvin, Lord (Thomson, W.) (1880) London Edinburgh Dublin Phil. Mag. J. Sci.,Fifth series, 33: 485-512.
6. Batchelor, G.K. (1950) Proc. Roy. Soc. London, A201: 405-416.
7. Majda, A.J. and Kramer, P.R. (1999) Phys. Rep., 314: 237-574.
8. Shraiman, B.I. and Siggia, E.D (2000) Nature, 405: 639-645.
9. Tennekes, H. and Lumley, J.L. (1972) A first course of turbulence, MIT Press, Massachusetts.
10. Chernyshenko, S.I. and Baig, M.F. (2003) The mechanism of streak formationin near-wall turbulence, J. Fluid Mech., sub judice.
11. Antonia, R.A., Ould-Rouis, M. Anselmet, F. and Zhu, Y. (1997) J. Fluid Mech., 332: 395–409.
12. Lüthi, B., Tsinober, A. and Kinzelbach, W. (2003) Lagrangian measurement of vorticity dynamics in turbulent flow, J. Fluid Mech., sub judice.
13. Tsinober A. and Galanti, B. (2003) Phys. Fluids, 15: 3514-3531.

Upper Bound on the Heat Transport in a Heated From Below Fluid Layer

Nikolay K. Vitanov[1,2]

[1] Institute of Mechanics, Bulgarian Academy of Sciences, Akad. G. Bonchev Str., Bl. 4, 1113, Sofia, Bulgaria vitanov@imbm.bas.bg

[2] Max Planck Institute for the Physics of Complex Systems, Nöthnitzerstr. 38, 01187, Dresden, Germany vitanov@mpipks-dresden.mpg.de

Summary. By means of the Howard-Busse method of the optimum theory of turbulence we obtain numerical upper bound on the Nusselt number in a horizontal heated from below layer of fluid of finite Prandtl number. We show that for low and intermediate Rayleigh numbers the numerical bound is below the analytical bound obtained by Howard. For large Rayleigh numbers the numerical bound approaches the analytical asymptotic bound from below.

Turbulent flows are highly chaotic and disordered dissipative nonlinear flows with many degrees of freedom and even the significantly increased computational power today do not allow us to perform direct numerical simulation of industrial and geophysical flows at large Reynolds and Rayleigh numbers. Thus other methods are also in use and very popular ones are the methods based on closure schemes. The optimum theory of turbulence avoids the closure problem by exploiting a different philosophy. Instead of dealing directly with the Navier-Stokes equations we can construct a variational problem on the basis of several integral relationships derived from these equations. The Euler-Lagrange equations of this variational problem lead to upper bounds on some quantities of interest such as the convective heat transport. If we add additional constraints in the variational problem we can tighten the bounds [1].

Howard and Busse [2, 3] derived the first upper bounds on the transport of heat by convection through a fluid layer by means of single-wavenumber and multi-wavenumber solutions of the Euler-Lagrange equations of the variational problem. Each of these solutions leads to upper bound on the convective heat transport in some interval of Rayleigh numbers. For applications of the Howard-Busse method to other problems of thermal convection see for an example [4, 5, 6]. We note the successful attempt to lower the bounds by using three- dimensional optimum fields instead of one-dimensional ones [1] as well as the Doering-Constantin method [7, 8] which could lead to quick estimates of the bounds on turbulence quantities.

Let us consider a horizontally infinite layer of fluid heated from below and let us use the model equations of the Boussinesq approximation. The thickness of the layer d is unit for length and the time scale is d^2/κ where κ is the thermal diffusivity of the fluid. The fixed temperatures T_1 and T_2 at the upper and lower boundaries of fluid participate in the unit for length $(T_2 - T_1)/R$ where $R = (\gamma(T_2 - T_1)gd^3)/(\kappa\nu)$ is the Rayleigh number. g is the acceleration of gravity, ν is the kinematic viscosity, and γ is the coefficient of thermal expansion. Starting from the Boussinesq equations and denoting by $\overline{q}$ and $\langle q \rangle$ the averaging of q over the horizontal plane and over the fluid layer we can obtain two momentum equations (for more details see [9]). When the convection is under stationary conditions long after any external parameter has been changed and all horizontally averaged quantities are time independent, we can use the momentum equations for formulation of the following variational problem

Find the minimum $R(\mu, w, \theta)$ of the functional

$$\mathcal{R}(\mu, w, \theta) = \frac{\langle \mid \nabla\theta \mid^2\rangle\langle \mid \nabla\mathbf{u} \mid^2\rangle}{\langle w\theta\rangle^2} + \mu \frac{\langle(\overline{w\theta} - \langle w\theta\rangle)^2\rangle}{\langle w\theta\rangle^2} \tag{1}$$

among all vector fields $\mathbf{u} = (u, v, w)$ and scalar fields θ which satisfy the continuity equation and the rigid boundary conditions $w = \theta = \partial w/\partial z = 0$ at $z = 0, 1$.

The continuity equation is taken into account by means of poloidal-toroidal representation of a solenoidal vector field and the Euler-Lagrange equations of the variational problem are

$$\langle \mid \nabla_1\nabla^2\phi \mid^2\rangle\nabla^2\theta + w[R\langle w\theta\rangle - \mu(\overline{w\theta} - \langle w\theta\rangle)] \tag{2}$$

$$\langle \mid \nabla\theta \mid^2\rangle\nabla_1^2\nabla^4\phi - \nabla_1^2\theta[R\langle w\theta\rangle - \mu(\overline{w\theta} - \langle w\theta\rangle)] = 0 \tag{3}$$

The homogeneity of the functional and the special kind of nonlinearity of the Euler-Lagrange equations allow us to impose the normalisation conditions: $\langle w\theta\rangle = \langle \mid \nabla \times \mathbf{k}\nabla^2\phi \mid^2\rangle = \mu$=R . Nu; $\langle \mid \nabla\theta \mid^2\rangle = \gamma\mu$, and to seek N-wavenumbers solutions of following form: $\phi = \sum_{n=1}^{N} \phi_n(x,y)\alpha_n^{-2}w_n(z)$, $\theta = \sum_{n=1}^{N} \phi_n(x,y)\theta_n(z)$, where $\nabla_1^2\phi_n = -\alpha_n^2\phi_n$, $\overline{\phi_n\phi_m} = \delta_{nm}$, δ_{nm} is the Kronecker delta symbol, $w_n(z) = \sum_{m=1}^{M} a_{nm}[\cos(2(m-1)\pi z) - cos(2m\pi z)]$, and $\theta_n(z) = \sum_{m=1}^{M} b_{nm}\sin[(2m-1)\pi z]$. These solutions satisfy the boundary conditions and the infinite sums must be truncated at sufficiently large value of the truncation parameter M in order to obtain finite system of nonlinear algebraic equations.

The results of the numerical investigations are presented in Fig. 1. Panel (a) presents the optimum field $w_1(z)$ for the $1-\alpha$-solution of the Euler- Lagrange equations of the variational problem for different values of Rayleigh number. We observe a formation of a boundary layer. In addition two peaks develop

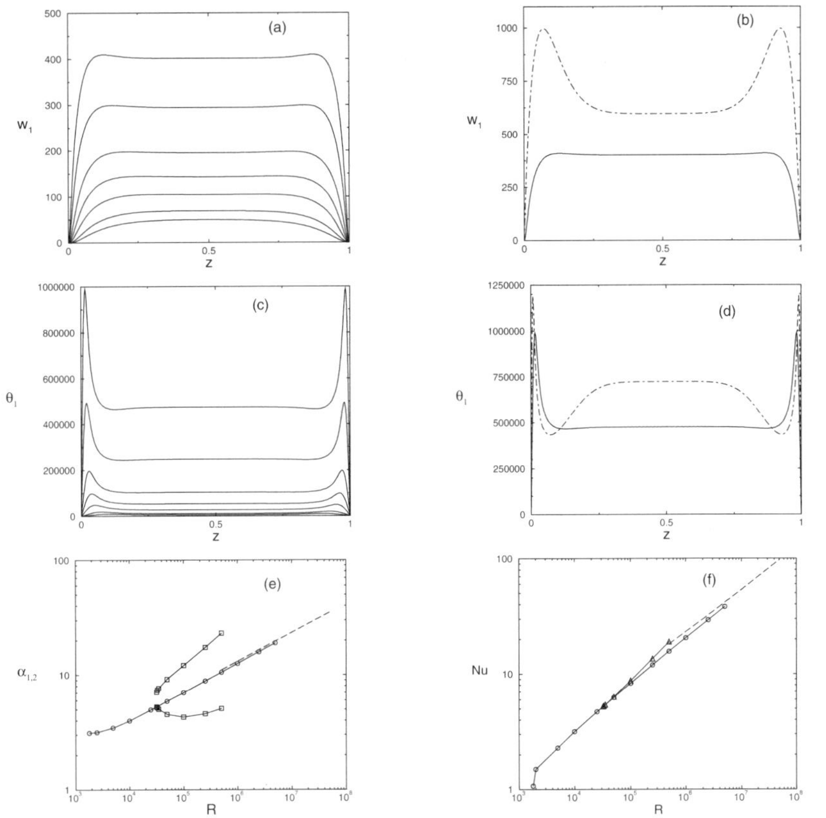

Fig. 1. Panels (a) and (c): Optimum fields w_1 and θ_1 for the one-wavenumber solution of the Euler-Lagrange equations for the case of fluid layer with rigid boundaries. From bottom to the top the profiles are plotted for the following values of Rayleigh number: $5 \cdot 10^4$, 10^5, $2.5 \cdot 10^5$, $5 \cdot 10^5$, 10^6, $2.5 \cdot 10^6$, $5 \cdot 10^6$. Panels (b) and (d): Comparison between optimum fields for the cases of rigid and stress-free boundaries. $R = 5 \cdot 10^6$. Solid lines: Rigid boundaries. Dot-dashed lines: Stress-free boundaries. Panel (e): Behavior of the wavenumbers $\alpha_{1,2}$. Dashed line: Asymptotic result of Howard [2] for the single wavenumber solution of the Euler-Lagrange equations. Circles: numerical results for the same case. Squares: $\alpha_{1,2}$ for the two-wavenumbers solution; α_1 (below the circles) and α_2 (above the circles). Panel (f): Numerical and analytical upper bounds $Nu(R)$ on convective heat transport. Dashed line: asymptotic result of Howard [2]. Circles: Numerical result for the same case. Triangles: upper bound given by the two-wavenumbers solution of the Euler-Lagrange equations.

slowly in the the function $w_1(z)$. These peaks are smaller in comparison to the corresponding peaks for the case of stress-free boundary conditions -see panels (b) and (d) for comparison between the two kinds of boundary conditions . In addition an internal layer of almost constant w_1 develops and its size increases with increasing Rayleigh number. An intermediate layer is formed between the boundary layer and the intermediate layer. The development of such layers and their behavior with increasing Rayleigh number are events supporting the assumptions made in the analytical asymptotic theory. The field $\theta_1(z)$ for the $1-\alpha$-solution of the Euler-Lagrange equations of the variational problem develops peaks very fast - panel (c) of Fig.1. The formation of internal, intermediate, and boundary layers is much faster in comparison to the formation of these layers for the field $\omega_1(z)$. The formation of peaks and layers in $\theta_1(z)$ proceeds approximately with the same tempo as the formation of these structures for the case of fluid layer with stress-free boundaries -see panel (d). This is despite the fact that $\theta_1(z)$ for the rigid boundaries case has smaller values in comparison to the same field for the case of stress-free boundaries. We note that at the highest Rayleigh number for which numerical results have been obtained the wave number of the $1-\alpha$-solution begins to reach from below its analytical asymptotic value $\alpha_1 = 0.417R^{1/4}$ as obtained by Howard -panel (e). Thus this is one direct numerical evidence for the validity of the analytical asymptotic theory of Howard. The Euler-Lagrange equation of the variational problem capture correctly also the critical Rayleigh number for the onset of convection. After this Rayleigh number the single-wavenumber solution of the Euler-Lagrange equations gives upper bound on the heat transport for a decade of Rayleigh numbers. At higher Rayleigh this upper bound reaches from below the upper bound $Nu = 1 + 0.125R^{3/8}$ obtained by Howard. Thus the analytical asymptotic results for the upper bound obtained by Howard are confirmed. Finally panels (e) and (f) of Fig. 1 show also the development of the optimum values $\alpha_{1,2}$ and the upper bound $Nu(R)$ for the two-wavenumber solution of the Euler-Lagrange equations. When calculated for large enough Rayleigh numbers these quantities can be used to test the multi-wavenumber theory of Busse [3].

The author thanks to Alexander von Humboldt Foundation for support of his research, to Prof. F. H. Busse for stimulating discussions and to NATO for the support through its Cooperative Science and Technology Sub-Program Grant No. PST.CLG.979126.

References

1. Vitanov N. K., Busse F. H (2001) Phys. Rev. E 63: 016303
2. Howard L. N. (1963) J. Fluid Mech. 17: 405–432
3. Busse F. H. (1969) J. Fluid. Mech. 37: 457–477
4. Vitanov N. K. (1998) Phys. Lett. A 248: 338–346
5. Vitanov N. K. (2000) Physica D 236: 322–339
6. Vitanov N. K. (2003) Phys. Rev. E 67: 026322
7. Doering C. R., Constantin. P. (1992) Phys. Rev. Lett. 69: 1648–1651
8. Kerswell R. R. (1998) Physica D 121: 175–192
9. Vitanov N. K., Busse F. H. (1997) ZAMP. 48: 310–324

Part II

Scaling Laws and Intermittency

Symmetries and Boundary Layer Profiles for Scalar Fields

Jahanshah Davoudi and Bruno Eckhardt

Fachbereich Physik, Philipps Universität, D-35032 Marburg, Germany
j.davoudi@physik.uni-marburg.de
bruno.eckhardt@physik.uni-marburg.de

1 Introduction

The "law of the wall" for the inner part of a turbulent shear flow over a solid surface is one of the cornerstones of fluid dynamics, and one of the very few pieces of turbulence theory where a simple analytic function for the mean profile of the velocity or temperature can be given [1, 2, 3]. While there seems to be ample experimental evidence for the logarithmic law of the wall there continues to be a heated debate over the validity [4], about the numerical factors that enter [5] and the regions for which it applies [6]. Some insight into the origin of the law and possible alternatives can be obtained by a Lie group analysis of the Navier-Stokes equations, as initiated by Oberlack [7, 8]. The important conclusions from such an analysis are the identification of alternative forms and the relation between continuous symmetries and the functional form. Mean profiles that have been obtained include an algebraic law in the center of a channel flow and in a near wall region, the viscous sub-layer, the linear mean velocity in the center of a Couette flow, the linear mean velocity in the center of a rotating channel flow, and an exponential mean velocity profile for the outer part of a boundary layer flow over a flat plate. The symmetry analysis elucidates the relation of these laws to symmetries so that the various mean profiles can be motivated on the assumption that the preferred state of the turbulent flow is the one with the highest degree of symmetry compatible with boundary conditions and other constraints. It is our aim here to summarize the results from a symmetry analysis of the various relations for the advection of a passive scalar in a turbulent flow [9].

2 Symmetry Analysis

A scalar field $\Theta(\mathbf{x}, t)$ transported by a velocity field $\mathbf{v}$ evolves in time according to the advection-diffusion equation

$$(\partial_t + \mathbf{v} \cdot \nabla)\Theta(\mathbf{x}, t) = \kappa \nabla^2 \Theta(\mathbf{x}, t). \tag{1}$$

For a passive scalar the dynamics of the velocity field $\mathbf{v}$ is not influenced by Θ and is given by the Navier-Stokes (NS) equation,

$$\partial_t \mathbf{v} + \mathbf{v} \cdot \nabla \mathbf{v} = -\nabla p + \nu \nabla^2 \mathbf{v}, \tag{2}$$

for an incompressible ($\nabla \cdot \mathbf{v} = 0$) fluid with kinematic viscosity ν and diffusivity κ. We assume that in the velocity field there is a mean large scale shear $\mathbf{U} = U(x_2)\mathbf{e}_1$, with x_1 coordinate in the flow-direction, x_2 the one in the direction of the shear and x_3 the one in the spanwise direction. We then decompose the velocity field $\mathbf{v} = \mathbf{U} + \mathbf{u}$, with $\mathbf{u}(\mathbf{x}, t)$ the turbulent velocity fluctuations. Similarly, we decompose the scalar field into a mean part $T(x_2)$ and fluctuations θ, such that $\Theta(\mathbf{x}, t) = T(x_2) + \theta(\mathbf{x}, t)$. Note that the dependence of the mean scalar profile on the wall normal distance x_2 is consistent with the moving boundary assumption. For stationary mean profiles ($\partial U/\partial t = 0$ and $\partial T/\partial t = 0$) the Reynolds averaged equations become [2, 1]

$$\begin{aligned} \mathcal{N}_i \equiv \frac{\partial u_i}{\partial t} + U\frac{\partial u_1}{\partial x_1} + \delta_{i,1} u_2 \frac{\partial U}{\partial x_2} - \delta_{i,1}(\gamma + \nu\frac{\partial^2 U}{\partial x_2^2}) + \delta_{i,2}\frac{\partial \bar{p}}{\partial x_2} \\ + \frac{\partial u_i u_k}{\partial x_k} + \frac{\partial p}{\partial x_i} = 0, \end{aligned} \tag{3}$$

$$\mathcal{K} \equiv \frac{\partial u_i}{\partial x_i} = 0, \tag{4}$$

$$\mathcal{L} \equiv (\frac{\partial}{\partial t} + U\frac{\partial}{\partial x_1})\theta + u_i\frac{\partial \theta}{\partial x_i} - \kappa\frac{\partial^2 T}{\partial {x_2}^2} + u_2\frac{\partial T}{\partial x_2} - \kappa\nabla^2\theta = 0, \tag{5}$$

where γ is the mean pressure applied in the streamwise direction x_1. The corresponding decomposed continuity equation for U is trivially satisfied. Following [7] we assume that the mean profiles adjust themselves so as to be invariant under a maximum number of symmetries compatible with the boundary conditions and external constraints. Moving across the boundary layer the constraints and thus also the symmetries and the invariant profiles may vary.

3 Symmetry Profiles

The search for all possible continuous symmetries is considerably assisted by tools developed within Lie's theory of continuous symmetries for smooth dynamical systems described by vector fields [10]. Given the relevant symmetry generators one may construct invariant manifolds of the set of dependent and independent variables. Technically, they are obtained by integrating the characteristic equations [10] of the infinitesimal generators. Under conditions of stationarity the characteristic equations can be rewritten [7, 10] in the form of differential equations for the mean profiles as a function of distance from the wall,

Table 1. Symmetry parameters and their physical meaning. Absence of a symmetry implies that the corresponding parameter is set to zero.

Group parameters	Physical meaning
a_1	scaling of position
a_4	scaling of time
a_3	translation in space
a_5	translation in time
b_1	translation in velocity
α	scaling of temperature
β	shift in temperature

$$\frac{dU}{dx_2} = \frac{(a_1 - a_4)U + b_1}{a_1 x_2 + a_3}, \tag{6}$$

$$\frac{dT}{dx_2} = \frac{\alpha T + \beta}{a_1 x_2 + a_3}. \tag{7}$$

According to Table 1 each of the parameters appearing in the above differential equation is attributed to one of the invariant groups of eqs. (3), (4) and eq.(5). When symmetries are broken the corresponding group parameter is put to zero. Translation symmetry in time is connected with stationarity and is always present, whence $a_5 \neq 0$. Translation symmetry in space can be broken near surfaces. Using the independence of the passive scalar dynamics and building on the previous analysis for the turbulent boundary layer profiles there are four cases for the velocity field, which can then be combined with two possible parameter combinations for the temperature field, giving a total of 8 combinations. The four cases for the velocity field are the well known logarithmic profile, an exponential profile, a linear profile and a power law profile. Similarly, there are linear, logarithmic, exponential and power law mean profiles for the passive scalar. The symmetries can be broken due to external scales in velocity, temperature or a prescribed length. The group parameter corresponding to that broken symmetry then vanishes.

We obtain velocity and temperature profiles that are logarithmic for $a_1 = a_4$ or $\alpha = 0$, and exponential for $a_1 = 0$. In case none of the group parameters vanishes both profiles are algebraic with an exponent that depends on the parameters; for $a_1 = a_4 = 0$ and $\alpha = 0$ linear laws result.

4 Profiles Across a Boundary Layer

Moving across a boundary layer several regions with different symmetries and hence different scaling laws are encountered. Consider, as an example, the case of $Pr = \nu/\kappa \simeq 0(1)$. On the smallest scales we have the thickness of the viscous boundary layer, $\delta_\nu = \nu/u_*$, and of the thermal boundary layer, $\delta_\kappa = \kappa/u_*$ [1, 2, 3]. For density $\rho = 1$, the friction velocity u_* is related to the shear stress τ at the surface by $u_* = \sqrt{\tau}$. On larger scales we have the overall thickness Δ of the dynamic and Δ_T of the thermal boundary layers. When the momentum and thermal turbulent boundary layers [1, 2, 3] are developed and the Prandtl number is of order one, then $\delta_\nu \simeq \delta_\kappa$ and $\Delta \simeq \Delta_T$.

In a first regime, close to the wall, with $x_2 \ll \delta_\nu \simeq \delta_\kappa$, the spatial scaling and velocity and temperature scaling symmetries are broken and the friction velocity u_* and temperature $T_* = Q/u_*$, with Q the total heat flux across the layer set the scales in that region. Both temperature and velocity increase linearly. In a second region, further out, for $x_2 \gg \delta_\nu$, the spatial scaling symmetry is preserved while the mean velocity and scalar scales are still determined by u_* and T_*. It is in this regime then that a logarithmic profile develops. Moving further out into the outer region of the boundary layer the thickness of thermal and velocity boundary layers again break the spatial scaling symmetry while the scalar and velocity are scale free. Then an exponential approach to the externally imposed mean values can be expected. Between the logarithmic and exponential region, another regime where the spatial symmetries and the rescaling symmetries of velocity and temperature are realized. This is the region in which the maximum number of scaling symmetries are preserved and where an algebraic law can be expected. Comparison to experimental data indicates that an identification of such a regime may be difficult.

Further discussions of the various scaling regions will be given in [9].

Most experimental measurements deal with the logarithmic profile, both in velocity and in temperature [3]. For the velocity field further scaling laws have been determined [7, 8]. An experimental test of the profiles further out would be most welcome.

This work was supported by the Deutsche Forschungsgemeinschaft.

References

1. Tennekes H and Lumley JL (1972) A first course in turbulence, MIT Press, Cambridge
2. Monin AS and Yaglom AM (1975) Statistical fluid mechanics I, II, MIT Press, Cambridge
3. Antonia RA and Danh HQ (1977) Phys. Fluids 20:1050–1057
4. Barenblatt GI (1993), J Fluid Mech 248:513–520
5. Durst F, Zanoun ES and Nagib H (2004) in IUTAM Symposium on Reynolds Number Scaling in Turbulent Flow, Smits AJ (ed), Kluwer, Dordrecht, pp 129–136.
6. Wosnik M, Castillo L and George W (2000), J Fluid Mech 421:115–145
7. Oberlack M (2000) ZAMM 80:791–800
8. Oberlack M (2001) J Fluid Mech 427:299–328
9. Davoudi J and Eckhardt B (2004), in preparation
10. Bluman GW and Kumei S (1989) Symmetries and Differential Equations, Springer, Berlin.

Energy and Dissipation Balances in Rotating Flows

Bruno Eckhardt[1], Siegfried Grossmann[1], and Detlef Lohse[2]

[1] Fachbereich Physik, Philipps-Universität Marburg, 35032 Marburg, Germany
bruno.eckhardt, siegfried.grossmann@physik.uni-marburg.de

[2] Applied Physics, University of Twente, 7500 AE Enschede, The Netherlands,
lohse@tn.utwente.nl

1 Introduction

The global analysis of energy flow and dissipation combined with a plausible ansatz for the splitting of dissipation between bulk and boundary layers has met with remarkable success in the case of a fluid layer heated from below [1, 2, 3]. The system parameters are captured by the Prandtl number Pr and the Rayleigh number Ra. The fluid responds to this external driving by developing a velocity field that causes an increased vertical heat transport. The velocity field is measured with a dimensionless Reynolds number Re and the increase in heat transport by the Nusselt number Nu, the ratio of the actual heat transport to the laminar one. The arguments proposed in [1-3] lead to predictions for the relations $Nu(Ra, Pr)$ and $Re(Ra, Pr)$, most notably the prediction of several different scaling regimes. A huge body of experimental data can be captured by this analysis.

Attempts to transfer these ideas to turbulence in shear flows date back to [4]. It was possible to fit experimental data for the pressure gradient in pipe flow [5] and for the torque G in a Taylor-Couette flow [6, 7] with scaling relations of the form $G = c_1 Re^{3/2+5\xi/2} + c_2 Re^{2+3\xi}$, where the correction $\xi \approx -0.05$ for both flows. Dubrulle and Hersant [8] analyzed Taylor-Couette flow further, included the geometry factors and suggested expressions with logarithmic terms. More recent experimental data [9] in systems with smooth and rough walls indicated a scaling compatible with $\xi = 0$ and a dependence of the ratio c_2/c_1 on the roughness of the walls.

Ref. [8] recalls and expands on the long known similarity between thermal convection and centrifugally unstable flows (e.g. [10]). The relation cannot be exact, however, as the two flows have different numbers of independent fields. It is the aim of the present note to analyze the similarities and differences between buoyancy and centrifugally driven flows in a simple situation where the number of fields match and where no complications due to curved coordinates matter: 2-d rotating flows.

2 Flows in Confined Geometries

In the full 3-d geometries the Rayleigh-Bénard system is described by five scalar fields: three velocity components, a temperature field and the pressure. A rotating shear flow has only three velocity components and the pressure. However, if thermal convection is confined to a 2-dimensional plane, the number of fields reduces to four. Similarly, in the rotating case we take a mean flow in the azimuthal direction that varies radially and an azimuthally invariant transversal flow. The analogy then associates the temperature field in the Rayleigh-Bénard system with the azimuthal velocity field in the rotating flow. Similarly, the wind in thermal convection and the transverse velocities in the rotating flows are identified.

Consider first the Rayleigh-Bénard system with a velocity field in Boussinesq approximation, $\mathbf{u} = (u(x,z)\,, 0\,, w(x,z))$ and a temperature field $\Theta(x,z)$. The equations of motion are

$$\partial_t u + (\mathbf{u}\cdot\nabla)u = -\partial_x p + \nu\Delta u + \beta g\Theta \tag{1}$$

$$\partial_t w + (\mathbf{u}\cdot\nabla)w = -\partial_z p + \nu\Delta w \tag{2}$$

$$\partial_t \Theta + (\mathbf{u}\cdot\nabla)\Theta = \kappa\Delta\Theta \tag{3}$$

$$\partial_x u + \partial_z w = 0\,. \tag{4}$$

The coordinates are such that the velocities are confined to the $x-z$ plane and gravitation acts along the x-direction.

The temperature is governed by a linear equation and appears linearly in the buoyancy term in the equation for u. By rescaling, the temperature can be replaced by a field that has dimensions of velocity. As units we take the gap width d and the kinematic viscosity ν, so that velocities are measured in units of ν/d and times in units of d^2/ν.

The buoyancy term has dimensions of an acceleration. This suggests a field v of dimension velocity, related to Θ by $g\beta\Theta = v\nu/d^2$. With this the equations of motion become

$$\partial_t u + (\mathbf{u}\cdot\nabla)u = -\partial_x p + \Delta u + v \tag{5}$$

$$\partial_t w + (\mathbf{u}\cdot\nabla)w = -\partial_z p + \Delta w \tag{6}$$

$$\partial_t v + (\mathbf{u}\cdot\nabla)v = (\kappa/\nu)\Delta v \tag{7}$$

$$\partial_x u + \partial_z w = 0 \tag{8}$$

The unperturbed convective temperature profile is linear, $\theta(x) = \Delta(1 - x/d)$, with Δ the temperature by which the lower plate is warmer. This temperature gradient then defines a velocity difference in v of size $\delta v = (Ra/Pr)(\nu/d)$, with Rayleigh number $Ra = g\beta d^3\Delta/(\kappa\nu)$ and Prandtl number $Pr = \nu/\kappa$. The velocity field is given by $v = Ra\Theta\nu/(Pr\, d\, \Delta)$.

The next step is to derive the dissipation-current relations [11]. The energy density of the 2-component velocity field is $E = (u^2+w^2)/2$, and its dissipation

$\epsilon_u = \nu\langle(\partial_x u)^2+(\partial_x w)^2+(\partial_z u)^2+(\partial_z w)^2\rangle_V = \frac{\nu}{d^2}\langle uv\rangle_V$, where $\langle\ldots\rangle_V$ indicates an average over volume and time.

Averaging the equation for the temperature over a plane perpendicular to the surface highlights the conserved heat current across the layer, $J = \langle uv\rangle_A - \kappa\partial_x\langle v\rangle_A$,is constant, independent of height x. When averaged over the vertical distance, the current becomes $J = \langle uv\rangle_V + \kappa\delta v/d$ with δv the difference in v between bottom and top plate. Thus, $\langle uv\rangle_V = J - \kappa\delta v/d$. The dissipation in the temperature field becomes $\epsilon_v = \kappa\langle|\nabla v|^2\rangle_V = \kappa\int_{\partial V}(v\nabla)v d\mathbf{a} = (Ra\kappa/d^2)J$. The Nusselt number is the thermal transport in units of the laminar one, $J_l = \kappa\delta v/d$, so that $J = Nu\,\kappa\delta v/d$. Putting all terms together we obtain

$$\epsilon_u = \frac{\nu^3}{d^4}RaPr^{-2}(Nu-1) \tag{9}$$

$$\epsilon_v = \frac{\nu^3}{d^4}Ra^2Pr^{-3}Nu\,. \tag{10}$$

These relations resemble the ones for thermal convection, remembering that $v/(\nu d^{-1}) = Ra\,Pr^{-1}\Theta/\Delta$.

For the corresponding rotating flow we take cartesian coordinates where x points in radial directions, z along the axis of rotation and y the direction of the main flow. The analog of the temperature difference will be a difference in the velocity in the y-direction at boundaries in the x-direction. The spatial dependencies of the fields are limited to x and z. Then the equations for the three velocity components u, v and w in x-, y- and z-direction, respectively, in the presence of a rotation (rate Ω) become

$$\partial_t u + (\mathbf{u}\cdot\nabla)u = -\partial_x p + \nu\Delta u - \Omega v \tag{11}$$

$$\partial_t w + (\mathbf{u}\cdot\nabla)w = -\partial_z p + \nu\Delta w \tag{12}$$

$$\partial_t v + (\mathbf{u}\cdot\nabla)v = \nu\Delta v + \Omega u \tag{13}$$

$$\partial_x u + \partial_z w = 0 \tag{14}$$

The rotation enters through the Coriolis terms only, the centrifugal terms being absorbed in the pressure. The equations are almost identical to the ones given above for the 2-d Rayleigh-Bénard system, when the azimuthal component of the velocity field is identified with the temperature field. The only difference is that in contrast to the temperature, which drives the u-component only, the rotation also gives rise to a term in the v-equation. The analogy also shows that the transverse velocity fields u and w are equivalent to the wind in the Rayleigh-Bénard system.

The analog of the dissipation-current relations (9,10) follows immediately with the same steps as before. We assume that $v(x=0) = \delta v$ and $v(x=d) = 0$ and define an external Reynolds number $Re_0 = d\delta v/\nu$. Then there is an equation for the transport of momentum across the layer, $J = \langle uv\rangle_A - \nu\partial_x\langle v\rangle_A = \text{const.}$, one for the dissipation of the wind,

$$\epsilon_u = -\Omega \langle uv \rangle_V \tag{15}$$

and one for the dissipation in the mean shear profile,

$$\epsilon_v = \frac{\nu}{d^2} Re_0 J + \Omega \langle uv \rangle_V \, . \tag{16}$$

Assuming a situation as in Taylor-Couette with the inner cylinder rotating and the outer one at rest, the velocity in the Reynolds number and in the rate of rotation are the same, i.e. $\Omega = \delta v/d$ and $Re_0 = \delta v\, d/\nu = \Omega d^2/\nu$. In the laminar case the velocity field is linear and the momentum current becomes $J_0 = -(\nu^2/d^2)Re_0$. The equivalent of the Nusselt number is the momentum current J divided by the laminar one, $Nu = J/J_0$. In dimensionless units the relations become

$$\epsilon_u = \frac{\nu^3}{d^4} Re_0 (\Omega d^2/\nu)(Nu - 1) \tag{17}$$

$$\epsilon_v = \frac{\nu^3}{d^4} \left(Re_0^2 Nu - Re_0 (\Omega d^2/\nu)(Nu - 1) \right) \tag{18}$$

The total dissipation becomes

$$\epsilon_{tot} = \epsilon_u + \epsilon_v = \frac{\nu^3}{d^4} Re_0^2 Nu \tag{19}$$

and scales simply with Nu and Re_0.

The relations (9) and (10) for the thermal system and (17) and (19) for the rotating system have the same structure. For the evaluation of the scaling relations of the momentum transport Nu and the wind Re with the Reynolds number Re_0 of the imposed shear flow in the case of bulk dominated turbulence, we use the scaling $Nu \sim Re$ for the current, $\epsilon_{tot} = Re_0^2 Re$ for the total dissipation and $\epsilon_u \sim Re^3$ for the dissipation in the wind. The latter equation together with (17) implies that the wind scales with the external shear, $Re \sim Re_0$. This implies the value $\xi = 0$, in accord with the recent data [9].

This work was supported by the EU within the Network Non-ideal turbulence, HPRN-CT-2000-00162, and the Deutsche Forschungsgemeinschaft.

References

1. Grossmann S and Lohse D (2000) J Fluid Mech 407:27–56
2. Grossmann S and Lohse D (2001) Phys Rev Lett 86: 3316–3319
3. Grossmann S and Lohse D (2002) Phys Rev E 66: 016305(6 pages)
4. Eckhardt B, Grossmann S, and Lohse D (2000). Europ Phys J B 18:541–545
5. Zagarola MV and Smits AJ (1998) J Fluid Mech 373: 33–79
6. Lathrop DP, Fineberg J, and Swinney HL (1992) Phys Rev A 46:6390–6405
7. Lewis GS and Swinney HL (1999) Phys Rev E 59: 5457–5467
8. Dubrulle B and Hersant F (2002) Europ Phys J B 26:379–386
9. van den Berg TH, Doering CR, Lohse D and Lathrop DP (2003), Phys Rev E 68:036307 (4 pages)
10. Bradshaw P (1969) J Fluid Mech 36:177–191
11. Shraiman BI and Siggia ED (1990) Phys Rev A, 42:3650–3653

Turbulent Cascade with Intermittency in View of Fragmentation Universalities

Mikhael Gorokhovski[1] and Vladimir Saveliev[2]

[1] CORIA UMR 6614 CNRS University of Rouen, 76 801, Saint-Etienne du Rouvray, France gorokhovski@coria.fr

[2] Institute of Ionosphere, Academy of Science, Kamenskoe Plato, 480020 Almaty, Kasahkstan

1 Introduction to Fragmentation Under the Scaling Symmetry

Fragmentation plays an important role in a variety of physical, chemical, and geological processes. Although each individual action of fragmentation is a complex process, the number of these elementary actions is large. Then it is natural to abstract a simple "effective" scenario of fragmentation and to represent its essential features. One of the model is the fragmentation under the scaling symmetry: each breakup action reduces the typical length of fragments, $r \Rightarrow \alpha r$, by an independent random multiplier α ($0 < \alpha < 1$), which governed by the fragmentation intensity spectrum $q(\alpha)$, $\int_0^1 q(\alpha)d\alpha = 1$. This scenario has been proposed by Kolmogorov [1], when he considered the breakup of solid carbon particle. Describing the breakup as a random discrete process, Kolmogorov stated that at latest times, such a process leads to the log-normal distribution. The population balance in the case of fragmentation under the scaling symmetry evolves according to the following integro-differential equation [2]:

$$\frac{\partial f}{\partial t} = (\widehat{I}_+ - 1)\nu f \tag{1}$$

where $f(r,t)$ is the normalized distribution of size, ν is the breakup frequency,

$$\widehat{I}_+ f = \int_0^1 f\left(\frac{r}{\alpha}\right) q(\alpha) \frac{d\alpha}{\alpha} \tag{2}$$

is the operator of fragmentation. To fulfill the evolution of distribution with time, we consider $q(\alpha)$ to be different from delta function. The question is: How does the distribution $f(r,t)$ evolve? This question can not be completely answered since the solution of the evolution equation (1) requires knowledge of the spectrum $q(\alpha)$, which is principally unknown function. At the same time, the operator $\widehat{I}_+$ in (1) is invariant under the group of scaling transformations $(r \to \alpha r)$. Due to this symmetry, the evolution of the distribution $f(r,t)$ goes at least, through two intermediate asymptotics. Evaluating these intermediate asymptotics does not require knowledge of entire function $q(\alpha)$ - only its first two logarithmic moments and further only the ratio of these moments in the long-time limit, determine the behavior of the solution of (1). These two *universalities* are shown as follows.

2 First and Second Universalities

The asymptotic solution of (1) is [2]:

$$f(r,t\to\infty) = \frac{1}{R}\frac{1}{\sqrt{2\pi\left\langle \ln^2\alpha\right\rangle \nu t}}\exp\left(-\frac{\left\langle \ln\alpha\right\rangle^2}{2\left\langle \ln^2\alpha\right\rangle}\nu t\right)\times$$
$$\times\exp\left(-\frac{\left\langle \ln(r/R)\right\rangle^2}{2\left\langle \ln^2\alpha\right\rangle \nu t}\right)\left(\frac{R}{r}\right)^{1-\langle\ln\alpha\rangle/\left\langle \ln^2\alpha\right\rangle} \tag{3}$$

where R denotes the initial length scale. The expression (3) confirms the main result of Kolmogorov [1]: the long-time limit distribution is log-normal (*first universality* with two parameters, which are the first and the second logarithmic moments of the fragmentation intensity spectrum). Further, it is seen from (3), that as time progresses, the second exponential multiplier tends to unity while the first one is decreasing. This implies that the final behavior of the distribution is dominated by (*second universality*):

$$f(r,t\to\infty) \propto \left(\frac{1}{r}\right)^{1-\frac{\langle\ln\alpha\rangle}{\langle\ln^2\alpha\rangle}} \tag{4}$$

with *only one universal parameter* $\langle\ln\alpha\rangle / \left\langle \ln^2\alpha\right\rangle$.

2.1 First Universality and Fokker-Planck Equation

The emerging corollary from the first universality is that changing of higher moments $\left\langle \ln^k\alpha\right\rangle$, $k>2$ in equation (1) does not affect its solution at times sufficiently larger than the life time of the breaking fragment. Consequently, the evolution equation (1) can be reduced *exactly* to the Fokker-Planck equation:

$$\frac{\partial f(r)}{\partial t} = \left[-\frac{\partial}{\partial r}r\left\langle \ln\alpha\right\rangle + \frac{1}{2!}\frac{\partial}{\partial r}r\frac{\partial}{\partial r}r\left\langle \ln^2\alpha\right\rangle\right]\nu f(r) \tag{5}$$

The solution of (5) verifies to be:

$$f(r,t) = \frac{1}{r}\int_0^{\infty} \frac{1}{\sqrt{2\pi\left\langle \ln^2\alpha\right\rangle \nu t}} \exp\left[-\frac{\left(\ln\frac{r_0}{r} + \left\langle \ln\alpha\right\rangle \nu t\right)^2}{2\left\langle \ln^2\alpha\right\rangle \nu t}\right] f_0(r_0)dr_0 \qquad (6)$$

where $f_0(r_0)$ is the initial distribution.

2.2 Second Universality, Fractals and Boltzman Distribution

The power distribution (4) implies the fractal properties of formed fragments in the long-time limit. Setting in (4) $x = \ln r$, one yields:

$$\Phi(x) = r\cdot f(r,t) \propto e^{-\frac{x}{h}} \;\; where \;\; h = -\frac{\left\langle \ln^2\alpha\right\rangle}{\left\langle \ln\alpha\right\rangle} \qquad (7)$$

From (7), one can see that the asymptotic power (*fractal*) distribution (4) plays the same role as the *Boltzmann* distribution in problems of statistical physics. This gives an idea to identify the parameter $\left\langle \ln\alpha\right\rangle / \left\langle \ln^2\alpha\right\rangle$ by making analogy with theory of Einstein on the Brownian motion. One gets:

$$\frac{\left\langle \ln^2\alpha\right\rangle}{\left\langle \ln\alpha\right\rangle} = \ln\left(\frac{r_*}{r_0}\right) \qquad (8)$$

where the typical length scale r_* characterizes a measure of the collective fracture resistance of material to the process causing the fragmentation.

3 Application to the Turbulent Cascade with Intermittency

The cascade in the isotropic turbulence with intermittency may also be viewed in the framework of fragmentation under the scaling symmetry. Here, the energy of larger unstable eddies is transferred to smaller one at the fluctuating rate. Since the number of degrees of freedom to produce each turbulent structure is very large, it is clear that controlling of each elementary breakup of eddy is a difficult task. The very simple way is again, to assume that at each repetitive step of cascade, the probability to find the velocity scale of a "daughter" eddy is independent of the velocity scale of its "mother" eddy; *i.e.* when the turbulent length scale r gets smaller, the velocity increment, $\delta_r v(x) = |v(x+r) - v(x)|$, is changed by independent positive random multiplier;

$$\delta_r v = \alpha\delta_l v, \text{ with } r \le l \qquad (9)$$

Such an expression can be found in papers of Castaing *et al.* [3, 4, 5, 6, 7]. These papers show: at progressively decreasing length scales, the PDF of $\delta_r v$

develops stretched tails with growing central peak. This implies that at small length scales, the small amplitude events alternate with events of strong velocity variation. The intermittency is manifested in the same way: highest strain in narrow (dissipative) regions of flow. By Taylor's hypothesis of "frozen turbulence", it is traditional to evaluate $\delta_r v$-statistics by one-point measurements. Consequently, the penetration towards smaller scale in the turbulent cascade may be viewed as evolution "in time" through $\tau_* = \ln(L_{int}/l_i)$, where L_{int} is integral length scale and l_i is eddy length scale [8]. Using this "time" and assuming Gaussian distribution at integral scale, equation (6) can be rewritten for the symmetric part of $\delta_r v$:

$$f(\delta_\tau v, \tau_*) = \frac{1}{\delta_\tau v} \int_0^\infty \frac{d(\delta_{int} v)}{\sqrt{2\pi \left\langle \ln^2 \alpha \right\rangle \tau_*}} \times$$

$$\exp\left[-\frac{\left(\ln \frac{\delta_{int} v}{\delta_\tau v} + \langle \ln \alpha \rangle \tau_* \right)^2}{2 \left\langle \ln^2 \alpha \right\rangle \tau_*} \right] G(\delta_{int} v) \tag{10}$$

The crucial problem in (10) is identification of the dimensionless cascade parameter $\left\langle \ln^2 \alpha \right\rangle / \langle \ln \alpha \rangle$. In view of discussion on (8), greater value of $\left\langle \ln^2 \alpha \right\rangle / \langle \ln \alpha \rangle$ corresponds to larger velocity increments that are involved into formation of smaller eddies. Such a "fluid resistance" to the process causing the cascade, may be characterized by the collective effect of viscosity. In literature, the length scale starting from which the viscous effects are manifested, is referred to as Taylor micro-scale. In this spirit: $\left\langle \ln^2 \alpha \right\rangle / \langle \ln \alpha \rangle = \ln(\lambda / L_{int})$ and $\langle \ln \alpha \rangle = const \cdot \ln(\lambda / L_{int})$ where λ is scale close to the Taylor micro-scale and *const* has to be adjusted by experiment. The evolution of (10), show the progressive non-Gaussianity with development of stretched tails. The computed *pdf's* from (10) reproduce measurements given by [9].

References

1. Kolmogorov A N (1941) Dokl Akad Nauk SSSR 2:99–101
2. Gorokhovski M, Saveliev V J (2003) J Physics of Fluids 15(1):184–192
3. Castaing B, Gagne Y, Hopfinger E J (1990) Physica D 46:177–200
4. Castaing B, Gagne Y Marchand M (1993) Physica D 68:387–400
5. Castaing B (1996) J Phys II France 6:105–114
6. Kahalerras H, Malecot Y, Gagne Y, Castaing B (1997) Phys Fluids 10(4):910–921
7. Naert A, Castaing B, Chabaud B, Hebral B, Peinke J (1998) Physica D 113:73–78
8. Friedrich R, Peinke J (1997) Phys Rev Lett 78(5):863-866
9. Mordant N, Metz P, Michel O, Pinton J-F (2001) Phys Rev Lett 87

Observational Impact of Surrogacy on the Turbulent Energy Cascade

Martin Greiner[1], Jochen Cleve[2,3], Jörg Schumacher[4], and Katepalli R. Sreenivasan[2]

[1] Corporate Technology, Information&Communications, Siemens AG, D-81730 München, Germany, martin.greiner@siemens.com

[2] ICTP, Strada Costiera, 11, 34014 Trieste, Italy, cleve@ictp.trieste.it, krs@ictp.trieste.it

[3] Institut für Theoretische Physik, Technische Universität, D-01062 Dresden, Germany

[4] Fachbereich Physik, Philipps-Universität, D-35032 Marburg, Germany, Joerg.Schumacher@physik.uni-marburg.de

The inertial-range dynamics of fully developed turbulence is often characterized in terms of structure functions $S_n(\Delta v_l) = \langle \Delta v_l^n \rangle \sim l^{\xi_n}$. The velocity increment $\Delta v_l = (\mathbf{v}(\mathbf{x}+\mathbf{l}) - \mathbf{v}(\mathbf{x})) \cdot \mathbf{l}/l$ is either taken as the component parallel (x) or perpendicular (y) to the mean flow direction. For scales $\eta \ll l \ll L$, where η is the dissipation scale and L the integral scale, the structure functions reveal power-laws with scaling exponents ξ_n. However, depending on the Reynolds number and the flow geometry, these scaling laws are either only approximate or come with a rather narrow scaling range.

A different way to extract scaling exponents is proposed by the Refined Similarity Hypothesis $\langle \Delta v_l^n \rangle \sim \langle \varepsilon_l^{n/3} \rangle l^{n/3}$, which relates structure functions to integral moments $\langle \varepsilon_l^n \rangle \sim l^{-\tau_n}$ of the energy dissipation and respective exponents via $\xi_n = n/3 - \tau_{n/3}$. The energy dissipation with viscosity ν is defined as

$$\varepsilon = \frac{\nu}{2} \sum_{i,j=1}^{3} \left(\frac{\partial v_i}{\partial x_j} + \frac{\partial v_j}{\partial x_i} \right)^2 \tag{1}$$

and requires full knowledge about all three velocity components. Since in experimental data only one, at most two components of the velocity field are accessible, various surrogate forms are constructed:

$$\varepsilon_{\mathrm{sur1}}(x) = 15\nu(\partial_x v_x(x))^2 \,, \tag{2}$$

$$\varepsilon_{\mathrm{sur2}}(x) = \frac{15}{2}\nu(\partial_x v_y(x))^2 \,, \tag{3}$$

$$\varepsilon_{\mathrm{sur3}}(x) = \frac{15}{4}\nu \left[2(\partial_x v_x(x))^2 + (\partial_x v_y(x))^2\right] \,. \tag{4}$$

Upon assuming isotropy, all three constructions have the same mean value as (1). Their coarse-grained amplitudes are determined via

$$\varepsilon_l(x) = \frac{1}{l} \int_{x-l/2}^{x+l/2} \varepsilon(x')dx' \; . \tag{5}$$

Fig. 1 illustrates the second-order moment based on (2) for an atmospheric boundary layer record [1]. Its Reynolds number based on the Taylor microscale $\lambda = \sqrt{\langle v_x^2 \rangle / \langle (\partial_x v_x)^2 \rangle}$ is $R_\lambda = 9000$, its estimated ratio between integral length and dissipation scale is $L/\eta = 5 \times 10^4$ and it comes with a longitudinal as well as a transverse velocity component. The logarithmic local slope of $\langle \varepsilon_l^2 \rangle$ turns out to be constant only in the upper part of the inertial range, where it is equal to $\tau_2 = 0.20$. The same outcome holds for the other two surrogate forms (3) and (4). For a turbulent flow with such a large Reynolds number this result is to some degree surprising and for the moment leaves open the question as to why the scaling range does not extend more into the intermediate inertial scale range.

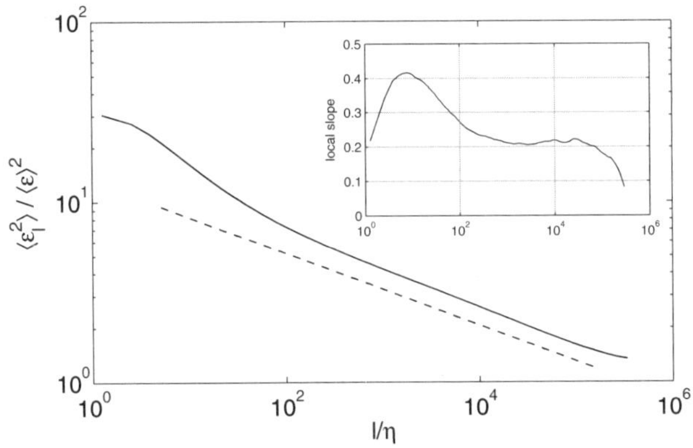

Fig. 1. Second-order integral moment $\langle \varepsilon_l^2 \rangle$ based on the surrogate energy dissipation (2) for an atmospheric boundary layer with $R_\lambda = 9000$. The dashed straight line has a logarithmic slope $\tau_2 = 0.2$. Inset shows the logarithmic local slope.

The second-order integral moment is closely related to the two-point correlation function:

$$\langle \varepsilon_l^2 \rangle = \frac{1}{l^2} \int_l dx_1 \int_l dx_2 \, \langle \varepsilon_{\text{sur}}(x_1) \varepsilon_{\text{sur}}(x_2) \rangle \; . \tag{6}$$

Fig. 2 compares the two-point correlator obtained from the surrogate forms (2), (3) and (4). All three variants reveal a rigorous power-law scaling behavior within the extended inertial range $15\eta \leq d \leq 0.3L$ and the corresponding scaling exponents are within $\tau_2 = 0.20 \pm 0.01$, showing little differences. Only for small two-point distances $d \to \eta$ the two-point correlators begin to differ. Whereas the variants based on (2) and (3) practically remain identical, the two-point correlation based on (4) is weaker for $d \leq 10\eta$; see inset of Fig. 2.

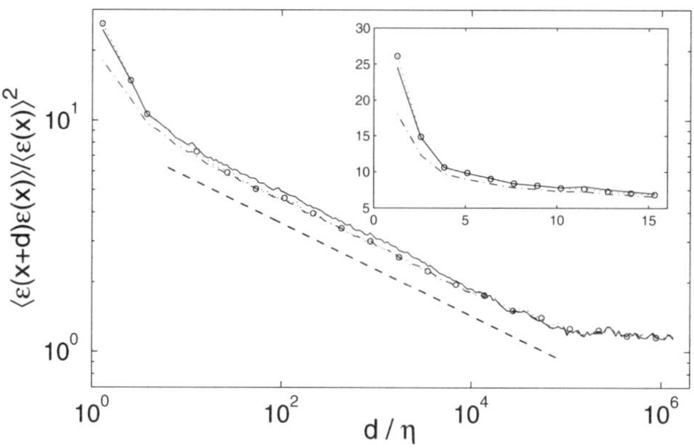

Fig. 2. Normalized two-point correlation function of the surrogate energy dissipation (2) (full line), (3) (dotted line with circles), and (4) (dot-dashed line). The dashed straight line has a logarithmic slope $\tau_2 = 0.2$. Inset magnifies the behavior for short separation distances.

When compared to the true energy dissipation, the expression (4) appears to be closer to (1) than the other two variants (2) and (3). This allows to speculate that if one adds more terms from the full list of (1), the extra-strong two-point correlations at small separation distances $d \leq 15\eta$ reduce further, perhaps even vanish once the surrogate field has converged to the true field.

In numerical simulations the full velocity field is accessible. Therefore they are particularly suitable for studying the difference between the true and surrogate energy dissipation. For the analysis here a small data set from a shear turbulence simulation [2] was available. Although the statistics are not very high the result is convincing enough to stress the surrogacy issue. Fig. 3 compares the two–point correlation of the dissipation obtained from the full field (1) with the one obtained from the surrogate field (2). Although the Taylor scale Reynolds number is only $R_\lambda = 99$ one can identify an approximate power law scaling range and both, the surrogate and true dissipation are identical in this range. Only for very small distances the two curves differ, where the correlator calculated from the surrogate field is showing the same extra-strong correlations as in experimental data. Note, that the strong increase of the correlation functions of the numerical data for the largest distances is an artifact of the periodic boundary conditions used in the simulation. This finding indicates the importance of the subtle surrogacy issue when interpreting data. The surrogacy of the energy dissipation alters the small scale behavior

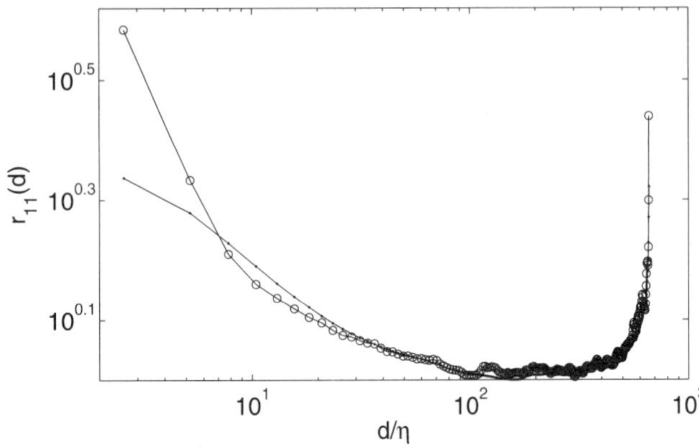

Fig. 3. Two-point correlator obtained from a numerical simulation. For the line with dots the true energy dissipation field (1) has been used and for the line with the open circles the one–component surrogate (2) has been used.

of the two-point correlation; fortunately, this leaves the rigorous scaling over the major part of the inertial range untouched.

In comparison with two-point correlations, the poor scaling of the integral moments appears in a new light. The extra-strong small-distance behavior of the two-point correlation can be roughly modeled with an additional δ-function at $d \approx \eta$, i.e. $\langle \varepsilon(x+d)\varepsilon(x) \rangle \approx a(\eta/d)^{\tau_2} + b\delta(d-\eta)$. Insertion into (6) then leads to $\langle \varepsilon_l^2 \rangle \approx a'(\eta/l)^{\tau_2} + b'(\eta/l)$, resulting in a slowly decreasing correction to the scaling term. This explains qualitatively the observed scale-dependence of the second-order integral moment: only in the upper part of the inertial range does the scaling term with exponent τ_2 dominate, whereas for the lower part strong deviations set in due to the small-distance behavior of the two-point correlation function, caused by the surrogacy effect.

More details on the observational impact of the surrogacy effect can be found in Ref. [3]. More follow-up discussions on two-point statistics of the turbulent energy cascade are given in Refs. [4, 5].

References

1. B. Dhruva, *An Experimental Study of High Reynolds Number Turbulence in the Atmosphere*, PhD thesis, Yale University (2000).
2. J. Schumacher J. Fluid Mech. **441**, 109 (2001).
3. J. Cleve, M. Greiner and K.R. Sreenivasan, Europhys. Lett. **61**, 756 (2003).
4. J. Cleve, M. Greiner and K.R. Sreenivasan, *On the intermittency exponent of the turbulent energy cascade*, arXiv:physics/0402015.
5. J. Cleve, Thomas Dziekan, Jürgen Schmiegel, Ole E. Barndorff-Nielsen, Bruce R. Pearson, K.R. Sreenivasan and M. Greiner, *Data-driven derivation of the turbulent energy cascade generator*, arXiv:physics/0312113.

Conditional Statistics of Velocity Increments in Fully Developped Turbulence

Antoine Naert

Laboratoire de Physique, École Normale Supérieure de Lyon, 46 Allée d'Italie, 69364 Lyon Cedex 07, France. Antoine.Naert@ens-lyon.fr

1 Introduction

The main shortcoming of Kolmogorov's theory [1] on the statistics of velocity fluctuations in high Reynolds-number flows is that, however beautifully simple, it does not take into account the inhomogeneity of intense velocity gradients. The presence of sparse but violent shear zones causing dissipation is referred to as *intermittency*. A signature of intermittency is the presence of large tails in the histograms of velocity gradients. Note that equivalently, the variance of velocity gradients is proportional to the dissipation rate: $\epsilon = \frac{1}{2}\nu(\partial_i v_j + \partial_j v_i)^2 \simeq 15\nu(dv/dx)^2$.
Obukhov and Kolmogorov proposed in 1962 an improved version of this model ($K.O.$ 62) to take into account spatial fluctuations of the energy transfer rate [2]. This is historically the first theory to address the scale-invariance symmetry breaking of the velocity field, that is intermittency.

This model is based on the rather strong assumption that scale-invariance would be recovered if only the regions of same value of the dissipation are taken into account in the ensemble averaging. This is the so-called *Refined Similarity Hypothesis* ($R.S.H.$).
The purpose of this study is to investigate experimentally those hypothesis underlying the theory, instead of the consequences.

2 Gaussian Conditional Statistics

First, we confirm with higher accuracy that the velocity increments are Gaussian at any inertial scale, if conditioned to the local average of the dissipation at the same scale. We use the traditional surrogate $\epsilon_r(x) = 15\nu \int_r \left(\frac{dv}{dx'}\right)^2 dx'$. This fact has already been observed since a long time by different groups [3, 4, 5]. The major improvement here is the use of a very clean data set recorded together with C. Baudet in a jet flow in air at moderate Reynolds

number ($R_\lambda \simeq 430$). All details about the experimental conditions and the resampling procedure can be found in [6].
One can see on fig.1 that the conditional histograms $P_c(v_r|\epsilon_r)$ are indeed Gaussian to a very good approximation.

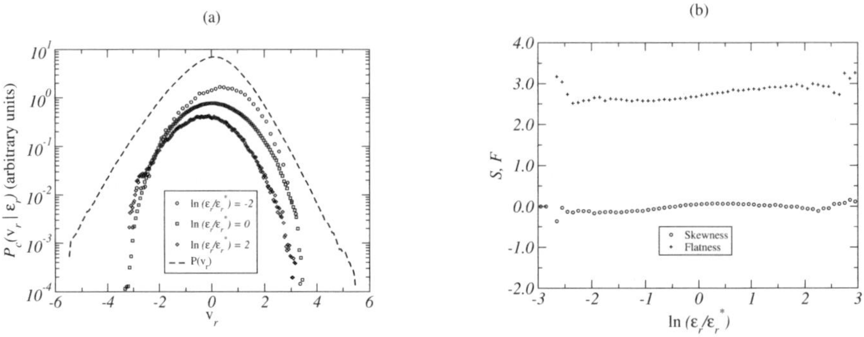

Fig. 1. (a) Conditional histograms $P_c(v_r|\epsilon_r)$, at scale $r = 100\,\eta$, for the following values of ϵ_r : $\log(\epsilon_r/\epsilon_r^*) = -2, 0, 2$, where ϵ_r^* denotes the most probable value of ϵ_r. The original histogram $P(v_r)$ at the same scale is plotted for comparison. Curves are artificially shifted along ordinates for clarity. (b) Skewness and flatness factors of the conditional histograms $P_c(v_r|\epsilon_r)$ vs. $\log(\epsilon_r/\epsilon_r^*)$ at the same scale $r = 100\,\eta$.

These histograms, for three different values of ϵ_r, have the same Gaussian shape in the whole inertial range (but not dissipative). Therefore, the following questions arise: – How do the 2 first conditional moments depend on r and ϵ_r, which is the *R.S.H.* itself? – What are the statistics of ϵ_r and the scale dependance of the variance (the mean being fixed by the energy conservation)? However, these last two question are not considered in this article: it is shown elsewhere that ϵ_r can be assumed log-normal to a very good approximation [8, 9, 10]. Several models give predictions on the variance of $log(\epsilon_r)$ [2, 7, 8].

3 Conditional Mean and Variance

$P_c(v_r|\epsilon_r)$ being Gaussian, are completely characterized by their two first moments: $v_0(\epsilon_r, r) \equiv \langle v_r|\epsilon_r\rangle$ and $\sigma^2(\epsilon_r, r) \equiv \langle v_r^2|\epsilon_r\rangle - \langle v_r|\epsilon_r\rangle^2$. Note that v_0 and σ characterize the asymmetry and the intermittency of v_r respectively.
One can see in fig. 2 that $v_0(\epsilon_r, r)$ and $\sigma(\epsilon_r, r)$ at different inertial scales all collapse to a single curve, if rescaled *à la Kolmogorov*:

$$\begin{cases} v_0(\epsilon_r, r) = \langle v_L^2\rangle^{1/2} \left(-\frac{1}{3}\,\log(\epsilon_r/\epsilon_r^*)\right)\,(r/L)^{1/3}\,, \\ \sigma(\epsilon_r, r) = \langle v_L^2\rangle^{1/2} \left(1 + \frac{1}{3}\,\log(\epsilon_r/\epsilon_r^*)\right)\,(r/L)^{1/3}\,. \end{cases} \quad (1)$$

Far from the most probable values of $\log(\epsilon_r)$ the conditional moments curve tends to zero, when the statistical sample becomes small. (The curvature is more visable for r large, where intermittency is small.)

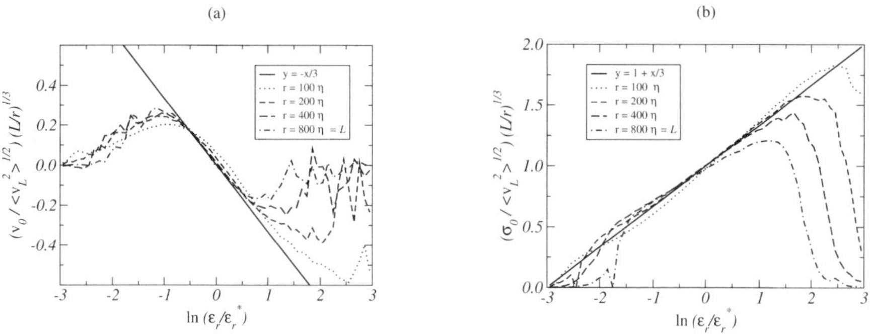

Fig. 2. A dimensional conditional mean velocity $(v_0(\epsilon_r, r)/\langle v_L^2\rangle^{1/2})(r/L)^{-1/3}$ (a) and conditional r.m.s. velocity $(\sigma(\epsilon_r, r)/\langle v_L^2\rangle^{1/2})(r/L)^{-1/3}$ (b) are plotted *v.s.* $\log(\epsilon_r/\epsilon_r^*)$ for several scales r in the inertial range and compared to the expressions given in Eq. (1).

4 Discussion

To conclude, we confirm with an increased accuracy that the conditional histogram $P_c(v_r|\epsilon_r)$ is Gaussian, and that its mean and standart-deviation scales as $r^{1/3}$, as predicted by *K.O.* 62. The two coefficients v_0 and σ describe respectively asymmetry and intermittency.

How can we interpret the fact that $P_c(v_r|\epsilon_r)$ is Gaussian?
It seems that this can only be understood from the Lagrangian point of view. The simplest Lagrangian model one can imagine for turbulence is that proposed by the Obukhov and Landau in 1944 [11]. A fluid particle is performing a Brownian motion, which expresses in the v-space as: $\langle v_\tau^2\rangle \propto \langle\epsilon\rangle\tau_v$. In this simplistic picture, v_τ is the Lagrangian velocity increment over a time τ_v, the mean dissipation $\langle\epsilon\rangle$ stands for *molecular diffusivity* which would be proportional to a *temperature* of the surrounding *reservoir*... if there were *thermodynamic equilibrium* of some sort.
At small scale, dissipation is fluctuating. To generalise this simplistic model, one can try to take into account these fluctuation as the *R.S.H.* for Eulerian turbulence: the random walk is performed in regions where dissipation (*temperature*) is not constant, giving rize to "anomalous diffusion". Therefore, one could recover "normal diffusion" characteristics in regions where ϵ_r takes the same value. This can be expressed as $\langle v_\tau^2|\epsilon_r\rangle \propto \epsilon_r\tau_v$, where ϵ_r is the average dissipation in the region of size r that the particle is exploring during time τ_v.

On this length r, ϵ_r must vary slower than v_τ, so that the particle can *thermalise*. In other words, we must compare the time $\tau_v \simeq r_v/v_\tau$ based on Lagrangian velocity, and the evolution time τ_ϵ of ϵ_r.
If we suppose that dissipation ϵ is passively advected by the turbulent field v we get $\tau_\epsilon \simeq r_\epsilon/v_\tau$, and find $r_\epsilon/r_v \simeq \tau_\epsilon/\tau_v$.
As correlations of ϵ and v are approximately algebraic and exponential respectively, it is likely that $r_\epsilon > r_v \Rightarrow \tau_\epsilon > \tau_v$. Therefore, the fluid particle has time to *thermalise*, and therefore exhibits at fixed ϵ_r the Brownian properties: $\langle v_\tau^2|\epsilon_r\rangle \propto \epsilon_r\tau$, and $P_c(v_r|\epsilon_r)$ Gaussian.
The scales on which those arguments hold are limited on one side by the small scales, on which the condition that ϵ_r is passively advected holds (that is turbulent mixing dominates thermal diffusion), and on the other side by the integral scales (on which $\epsilon_r \simeq \langle\epsilon\rangle$). That is nothing but the inertial range of scales.

Introducing intermittency this way seems to extend nicely the simplistic model of 1944 in a similar way as the *R.S.H.* of 1962 did for the 1941 theory.
Those qualitative arguments are obviously not sufficient. A theory is badly needed, that would link the Lagrangian and Eulerian pictures of turbulence, and if possible explain the dependance in ϵ_r of the conditional moments v_0 and σ. That would give a description of both intermittency and asymmetry.

References

1. A.N. Kolmogorov. *Dokl. Akad. Nauk. SSSR.*, 30: 299, 1941.
2. A.M. Obukhov. *J. Fluid Mech.*, 13: 77, 1962,
   A.N. Kolmogorov. *J. Fluid Mech.*, 13: 82, 1962.
   L.-P. Wang *et al.* *J. Fluid Mech.*, 309:113, 1996, and references therein.
3. G. Stolovitsky, P. Kailasnath, K. R. Sreenivasan. *Phys. Rev. Lett.*, 69:1178, 1992.
4. Y. Gagne, M. Marchand, B. Castaing. *J. Phys. II (France)*, 4:1, 1994.
   M. Marchand, Doctoral thesis 1994, Univ. Grenoble, France, unpublished.
5. A. Naert Doctoral thesis 1995, Univ. Grenoble, France, unpublished.
   A. Naert, B. Castaing, B. Chabaud, B. Hébral, J. Peinke. *Physica D*, 113:73, 1998.
6. P. Marcq, A. Naert. *Phys. Fluids*, 13:2590, 2001.
7. B. Castaing, Y. Gagne and E. Hopfinger. *Physica D*, 46:177, 1990.
8. A. Naert, R. Friedrich and J. Peinke, Phys. Rev. E. **56**, 6719 (1997).
9. P. Marcq, A. Naert. *Physica D*, 124:368, 1998.
10. A. Arneodo, S. Manneville, J. F. Muzy, S. G. Roux, A. Arneodo *et al.* *Phil. Trans. R. Soc. Lond. A*, 357:2415, 1999. Revealing a log-normal cascading process in turbulent velocity statistics with wavelet analysis.
11. Landau, Lifshitz (1963), *Fluid mechanics*, Pergamon Press, London.

A Simple Relation Between Longitudinal and Transverse Increments

Malte Siefert, Joachim Peinke[1], and Rudolf Friedrich[2]

[1] Carl-von-Ossietzky University, Institute of Physics, 26111 Oldenburg, Germany
[2] Westfälische Wilhelms-Universität, Institute of Theoretical Physics, 48149 Münster, Germany

The understanding of the complex statistics of fully developed turbulence in detail is still an open problem. One of the central points is to understand intermittency, i.e. to find exceptionally strong fluctuations on small scales. In the last years, the intermittency in different directions has attracted considerable interest. It has been controversial whether there are significant differences in intermittency between the different directions. More specifically one looks at the statistics of increments $[\mathbf{u}(\mathbf{x}+\mathbf{r}) - \mathbf{u}(\mathbf{x})]\,\mathbf{e}$, i.e. at the projection of the differences between two velocities separated by the vector $\mathbf{r}$ in a certain direction $\mathbf{e}$. Here we denotes longitudinal increments with u, for which $\mathbf{r}$ and $\mathbf{e}$ are parallel and transverse increments with v for which $\mathbf{r}$ is perpendicular to $\mathbf{e}$.

In a first step, one commonly investigates the statistics with the moments of the increments, the so-called structure functions, and assumes that, according to Kolmogorov, the structure functions obey a scaling law $\langle u^n \rangle \propto r^{\xi_n}$ at least for sufficient high Reynolds number. The intermittency problem is then expressed by the deviation of the exponent ξ_n from the value $n/3$, the well-known Kolmogorov (1941) scaling.

1 Multi-scale Statistics

In this paper, we focus on a different approach which enables to characterize multi-scale statistics. Recently it has been shown that it is possible to get access to the joint probability distribution $p(u(r_1), u(r_2), \ldots, u(r_n))$ via a Fokker-Planck equation, which can be estimated directly from measured data [4, 3]. This method is more general than the above mentioned analysis by structure functions, which characterize only the simple scale statistics $p(u(r))$ or $p(v(r))$. The Fokker-Planck equation (here written for vector quantities) reads as

$$-r\frac{\partial}{\partial r}p(\mathbf{u},r|\mathbf{u_0},r_0) = \tag{1}$$
$$\left(-\sum_{i=1}^{n}\frac{\partial}{\partial u_i}D_i^{(1)}(\mathbf{u},r)+\sum_{i,j=1}^{n}\frac{\partial^2}{\partial u_i\partial u_j}D_{ij}^{(2)}(\mathbf{u},r)\right)p(\mathbf{u},r|\mathbf{u_0},r_0).$$

(i labels the components of $\mathbf{u}$, we fix $i = 1$ for the longitudinal and $i = 2$ for the transverse increments.) This representation of a stochastic process is different from the usual one: instead of the time t, the independent variable is the scale variable r. The minus sign appears from the development of the probability distribution from large to small scales. In this sense, this Fokker-Planck equation may be considered as an equation for the "dynamics" of the cascade, which describes how the increments evolve from large to small scales under the influence of deterministic ($D^{(1)}$) and noisy ($D^{(2)}$) forces. The whole equation is multiplied without loss of generality by r to get power laws for the moments in a more simple way. Both coefficients, the so-called drift term $D_i^{(1)}(\mathbf{u},r)$ and diffusion term $D_{ij}^{(2)}(\mathbf{u},r)$, can be estimated directly from the measured data using its mathematical definition. With the notation $\Delta u_i(r,\Delta r) := u_i(r-\Delta r) - u_i(r)$ the definitions read as:

$$D_i^{(1)}(\mathbf{u},r) = \lim_{\Delta r\to 0}\frac{r}{\Delta r}\langle\Delta u_i(r,\Delta r)\rangle|_{\mathbf{u}(r)}, \tag{2}$$

$$D_{ij}^{(2)}(\mathbf{u},r) = \lim_{\Delta r\to 0}\frac{r}{2\Delta r}\langle\Delta u_i(r,\Delta r)\Delta u_j(r,\Delta r)\rangle|_{\mathbf{u}(r)}. \tag{3}$$

Here we extend the analysis to a two-dimensional Markov process, relating the longitudinal and transverse velocity increments to each other. The resulting Fokker-Planck equation describes the joint probability distribution $p(u(r_1),v(r_1);\ldots;u(r_n),v(r_n))$.

2 The Experiment

The analysis is applied to a signal measured in a wake behind a cylinder with 2 cm diameter. The mean free-stream velocity is 10 m/s resulting in a Reynolds number of 13000. We have placed a X-wire 60 diameter behind the cylinder to measure the stream wise velocity component u and the component v perpendicular to the cylinder axis. Taylor's hypothesis of frozen turbulence is used to convert time lags into spatial displacements. With the sampling frequency of 25kHz the spatial resolution does not resolve the Kolmogorov length but the Taylor length $\lambda = 4.85$ mm, yielding the Taylor based Reynolds number R_λ=180. The integral length is $L = 137$ mm.

3 Longitudinal versus Transverse Increments

From the measured data the drift and diffusion coefficients are estimated according to eqs. (2) and (3) as described in [4]. To use the results in an analytical way, the drift and diffusion coefficient can be well approximated by the following low dimensional polynoms,

$$\begin{aligned} D_1^{(1)}(u,v,r) &= d_1^u(r)u, \;\; D_2^{(1)}(u,v,r) = d_2^v(r)v \\ D_{11}^{(2)}(u,v,r) &= d_{11}(r) + d_{11}^u(r)u + d_{11}^{uu}(r)u^2 + d_{11}^{vv}(r)v^2 \\ D_{22}^{(2)}(u,v,r) &= d_{22}(r) + d_{22}^u(r)u + d_{22}^{uu}(r)u^2 + d_{22}^{vv}(r)v^2 \end{aligned} \tag{4}$$

The coefficients $d_{11}^{uu}(r)$ and $d_{22}^{vv}(r)$ are responsible for intermittency, the coefficient $d_{11}^u(r)$ is essential for the skewness of the probability distribution (Kolmogorov's 4/5-law). The verification of these coefficients will be presented in a more detailed publication.

Comparing the d-coefficients of equation (4) one can see a remarkable difference of the longitudinal and transversal increments. The r-dependence of related longitudinal and transverse d-coefficients (d_1^u and d_2^v etc.) coincides if the abscissa are rescaled: $d_{\text{long}}(r) \approx d_{\text{transv}}(2/3r)$, e.g. $d_1^u(r) = d_2^v(2/3r)$ etc.. Only the coefficients which are responsible for the skewness and the coefficient which is responsible for the intermittency deviate from this behavior.

Next, we give three examples, which show the consistency of the 2/3 rescaling. First, plotting the corresponding structure functions with the 2/3 rescaling we find that the longitudinal and transverse structure functions become similar. Secondly, one can interpret the Kármán equation (isotropic relation) as an Taylor expansion

$$\langle v^2(r)\rangle = \langle u^2(r)\rangle + \frac{1}{2}r\frac{\partial}{\partial r}\langle u^2(r)\rangle \approx \langle u^2(r+\frac{1}{2}r)\rangle = \langle u^2(\frac{3}{2}r)\rangle \tag{5}$$

to see again a rescaling of the scale variable r by the factor 2/3. This first order Taylor expansion is correct within 4% over the whole inertia range. Thirdly, let us suppose that the structure functions scale with a power law, $\langle v^n(r)\rangle = c_t^n r^{\xi_t^n}$ and $\langle u^n(r)\rangle = c_l^n r^{\xi_l^n}$. We can relate the structure functions according to the above mentioned rescaling: $\langle v^n(r)\rangle = \langle u^n(\frac{3}{2}r)\rangle = c_t^n r^{\xi_t^n} = c_l^n(\frac{3}{2}r)^{\xi_l^n}$. We end up with the relation $\xi_l^n = \xi_t^n$ and $\frac{c_t^n}{c_l^n} = \left(\frac{3}{2}\right)^{\xi_l^n}$. Note that the c^n constants are related to the Kolmogorov constants. For $n = 2$ and 4 we obtain $c_t^2/c_l^2 \approx 1.33$ and $c_t^4/c_l^4 \approx 1.72$, which deviates less than 3% from the value of $c_t^2/c_l^2 = 4/3$ and $c_t^4/c_l^4 = 16/9$ given in [1].

4 Extended Self Similarity of Longitudinal and Transverse Increments

To study the intermittency in the framework of self-similarity, the scaling exponent has to be measured. For low Reynolds number along with a short scaling range this is possible by using the extended self-similarity method (ESS) proposed by Benzi *et al.* [2] with the relations $\langle u^n\rangle \propto \langle |u^3|\rangle^{\xi_{l,n}}$ and $\langle v^n\rangle \propto \langle |u^3|\rangle^{\xi_{t,n}}$. A frequently discussed result is that $\xi_{t,n} < \xi_{l,n}$. This has been interpreted as a higher intermittency of the transversal increments, for our data, see Fig. 1a). But if the transverse structure function is plotted as a function of $\langle|u^3(\frac{3}{2}r)|\rangle$, this discrepancy vanishes, as shown in Fig. 1b). Notice that these properties are only possible for a non-existing scaling behavior.

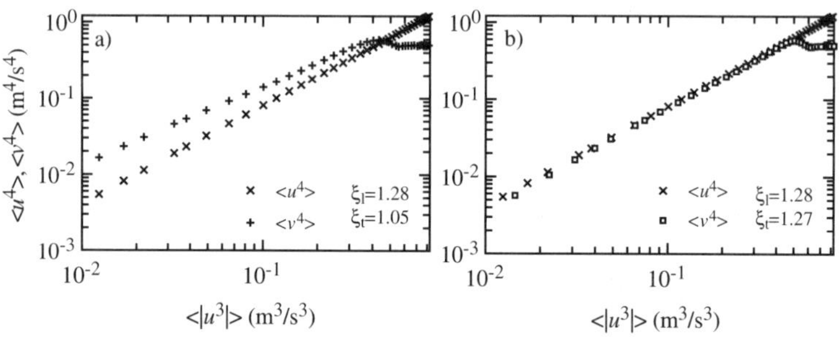

Fig. 1. Extended self similarity (ESS) for the forth-Worder structure function. a) ESS as it is usual applied for longitudinal and transverse structure functions. b) as in a) but the transverse structure function plotted against $\langle|u^3(\frac{3}{2}r)|\rangle$ with the abscissa stretched by the factor 2/3.

In conclusion, we have shown a new feature of the relation between the longitudinal and transverse velocity increments, namely, that the statistics of both quantities become similar if the relative "cascade speeds" between these two directions are rescaled by a factor 2/3. This finding seems to be more basic than the proposed scaling property of structure functions.

References

1. R. Antonia, M. Ould-Rouis, Y. Zhu, and F. Anselmet. Fourth-order moments of longitudinal- and transverse-velocity structure functions. *Europhys. Lett.*, 37:85–90, 1997.
2. R. Benzi, S. Ciliberto, R. Tripiccione, C. Baudet, F. Massaioli, and S. Succi. Extended self-similarity in turbulent flows. *Phys. Rev. E*, 48:R29–R32, 1993.
3. R. Friedrich and J. Peinke. Description of a turbulent cascade by a Fokker-Planck equation. *Phys. Rev. Lett.*, 78:863–866, 1997.
4. C. Renner, J. Peinke, and R. Friedrich. Experimental indications for Markov properties of small-scale turbulence. *J. Fluid Mech.*, 433:383–409, 2001.

Intermittency Exponent in High-Reynolds Number Turbulence

Yoshiyuki Tsuji

Department of Energy Engineering and Science, Nagoya University, Furo-cho, 464-8603, Nagoya, Japan c42406a@nucc.cc.nagoya-u.ac.jp

1 Introduction

According to Kolmogorov's similarity hypothesis (**K41**), the one-dimensional energy spectrum has the following form in the inertial range[1],

$$E(k) = C_k \langle\varepsilon\rangle^{2/3} k^{-5/3}, \tag{1}$$

where $\langle\varepsilon\rangle$ is the energy dissipation rate per unit mass and k is a wave number defined as $2\pi f/U$ by the frequency f[Hz]. However, this scaling is not satisfied completely. The spectrum indicates a power-law relation $E(k) \propto k^{-\gamma}$ whose slope is different from $-5/3$. When the power-law exponent is expressed by $\gamma \equiv -(5/3 + \mu)$, μ is a very small quantity but has been thought to be on the order of $O(10^{-2})$. The spectral slope has not been examined directly so far because the inertial range is too narrow to accurately measure the deviation from $-5/3$ in moderate laboratory experiment or DNS.

Recently, the ultra simulation of box turbulence was performed by EARTH SIMULATOR [2]. The Reynolds number based on the Taylor micro scale was $R_\lambda \simeq 1200$. They observed that the energy spectrum in the inertial subrange almost follows the Kolmogorov scaling law, but the exponent is steeper than $-5/3$ by about 0.1 within $0.008 \leq k\eta \leq 0.03$. If this was an intermittency correction μ, it would be too large. But no physical interpretation of this fact was presented by the authors. Where is this large discrepancy coming from? How can we interpret $\mu \simeq 0.1$? We address this question from an experimental point of view by analyzing the high-Reynolds number data measured in the atmospheric turbulent boundary layers [3].

2 Experimental Condition

The atmospheric data were measured in the observation tower at Brookhaven National Laboratory (see Fig. 1 in reference [3]). The station is located about 35 m from the ground. We used I-type and X-type probes whose sensitive lengths are 0.7 mm with diameters of 5 μm. They are operated by a constant-temperature anemometer. The velocity signals are sampled by 12-bit A/D converter at 5 kHz $\sim$ 10 kHz frequency with a low-pass filter depending on the flow conditions, and the measurements are continuous for one hour for each data set. The probe was mounted on a 2 m-long moving device, and its direction is arranged according to the flow condition. The Reynolds number is $5000 \leq R_\lambda \leq 30000$ and typical features of flow field are summarized in [3].

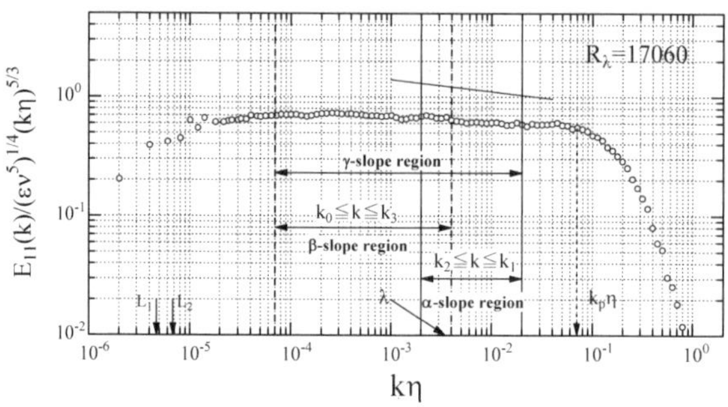

Fig. 1. Energy spectrum $E(k)$ normalized by the energy dissipation rate $\langle \varepsilon \rangle$ and wave number k. k_p is a peak wave number in the dissipation range. $k_0 \leq k \leq k_3$ and $k_2 \leq k \leq k_1$ are defined as β-slope and α-slope region, respectively. γ-slope region extends from k_0 to k_1. Here k_0 indicates the location where the normalized spectrum decreases in the low wave number range. Solid line indicates the slope $\mu = 0.1$.

Figure 1 shows an example of normalized spectra. η is defined by $\langle \varepsilon \rangle$ and viscosity ν as $\langle \varepsilon \rangle^{3/4} / \nu^{1/4}$. There are more than three decades of inertial range, and a fine resolution is achieved even in the dissipation range.

3 Power-Law Exponent of Energy Spectrum

It is actually observed that the energy spectrum is rather tilted in the range of $0.002 \leq k\eta \leq 0.02$. The solid line indicates the slope of $\mu = 0.1$ as observed by the ultra simulation. The flatter region, that is the spectral slope closer to $-5/3$, exists in the lower wave-number range. For convenience, the special wave numbers k_p, k_0, k_1, k_2, and k_3 are introduced. k_p is defined as the peak of the normalized spectrum in the dissipation range. It is representative of the small-scale fluctuations. If k_p is normalized by η, the ratio is constant, $k_p\eta \simeq 0.07$, in this data set. Toward the lower wave number range the spectrum decreases and then indicates the local minimum at k_1 where k_1 is approximately $k_1\eta = 0.02$. k_1 is located in the beginning of the bottleneck region. Toward the lower frequency range, the normalized spectrum deviates from the power-law around the wave number of k_0, and then the spectrum attenuates. We think that k_0 corresponds to the beginning of the inertial range or it is representative of large-scale fluctuations. It is noted that k_0 is not constant even if it is normalized by η. Then the ratio k_p/k_0 represents how far apart the small scale is apart from the large scale. We quantitatively characterize the wave numbers k_2 and k_3. They are not determined empirically but they are on the same order as the Taylor micro scale λ. For instance, λ is 12.8cm or $\lambda^{-1}\eta = 0.0035$ for the case of $R_\lambda = 17060$ in Fig. 1. That is, $k_2 \leq 1/\lambda \leq k_3$. An overlap region should exist due to some uncertainty in computing the spectrum. The spectral slope in $k_2 \leq k \leq k_1$ is expressed as α and that of $k_0 \leq k \leq k_3$ is β. We call them the α-slope region and β-slope region, respectively.

The slopes α and β are obtained by a least-square fit, and they are plotted as a function of k_p/k_0 in Fig. 2. As indicated by the dashed line, α is about 0.075 and $\beta = 0.025$ on average. β is actually smaller than α. Thus, the power-law region closer to $-5/3$ in the range of $k_0 \leq k \leq k_3$ is convincing. We assume that the α range corresponds to the power-law range observed in the ultra simulation with $\mu = 0.1$ [2]. α is slightly smaller than 0.1, but we think this is because the slope between k_1 and k_p is not as steep or the spectral bump is not as clearly observed in one-dimensional spectra (experiments) as in three-dimensional ones (DNS). This fact has been noticed, and Dobler et al. propose that this difference is caused by means of transforming the three-dimensional spectrum into one dimensional [4]. Once the one-dimensional spectrum is computed from the three-dimensional data, the spectral bump becomes less pronounced.

In ordinary laboratory experiments, R_λ is several hundreds at most, or the inertial range is about one decade. The spectrum indicates almost $-5/3$. This is commonly observed in DNS [5]. In this case, there is only the β-slope region but the α-region does not exist. Here we remember that the α-region was defined as the steep slope region where the deviation from $-5/3$ is the

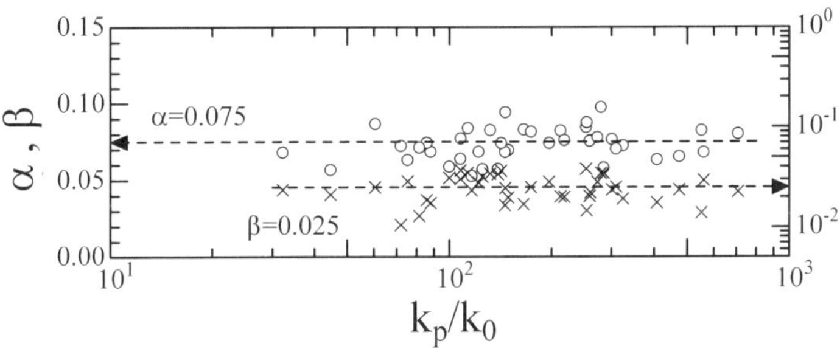

Fig. 2. Slope of the normalized spectrum in both α and β-regions. They are plotted as a function of the ratio k_p/k_0.

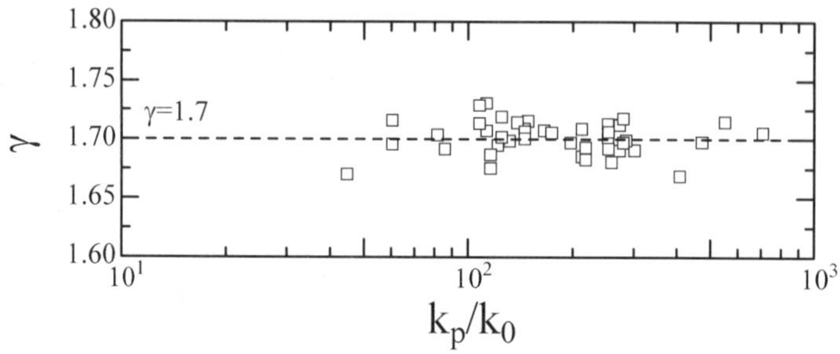

Fig. 3. Spectral slope computed by fitting the relation $E(k) \propto k^{-\gamma}$ in the range $k_0 \le k \le k_1$.

order of $O(0.1)$. As the Reynolds number increases so R_λ is a thousand, the β-region shifts into the low wave number range, and the α-slope region appears. This is the condition of ultra simulation at $R_\lambda \simeq 1200$.

Both α and β are thought to be an intermittency correction from **K41**. Near the spectral bump, $k_2 \le k \le k_1$, the intermittency correction is more significant. Thus, we have to take into account a different energy cascade process in these instances. If the inertial range exists far from the dissipation range in the sense of Kolmogorov, it should be regarded as a β-slope region. A substantial β-slope region is realized for $O(10^4) < R_\lambda$. The existence of two different scaling regions is interesting for itself, but we do not have enough knowledge to reveal the phenomenology in the α-region.

It is difficult to confirm the difference between α and β-slope regions without being motivated by the ultra simulation to look carefully into the energy spectrum. Usually, the spectral slope has been computed using the form $E(k) \propto k^{-\gamma}$ in the range of $k_0 \le k \le k_1$, or following the criterion that the normalized spectra $E(k)/k^{-\gamma}$ shows the broadest flat region. The exponent γ is plotted in Fig. 3. It is approximately 1.70 on average and independent of the ratio k_p/k_0. So the intermittency correction is $\gamma - 5/3 = 0.04$. More detailes are reported in [6].

References

1. K.R.Sreenivasan and R.A.Antonia, Annu. Rev. Fluid Mech., **29**, 435 (1997).
2. Y.Kaneda, T.Ishihara, M.Yokokawa, K.Itakura, and A.Uno, Phyics of Fluids, **15**, L21, (2003).
3. Y.Tsuji and B.Dhruva, Physics of Fluids, **11**, 3017, (1999).
4. W.Dobler, N.E.L.Haugen, T.A.Yousef, and A.Brandenburg, Physical Rev. E, **68**, 026304, (2003).
5. T.Gotoh, D.Fukayama, and T.Nakano, Physics of Fluids, **14**, 1065, (2002).
6. Y.Tsuji, accepted for publication in Physics of Fluids, (2004).

Part III

Modelling

An Alternative Model for Turbulent Flow and Forced Convection

Stuart W. Churchill

Department of Chemical and Biomolecular Engineering, University Of Pennsylvania 311A Towne Bldg., 22o South 33rd St., Philadelphia, Pennsylvania 19104, USA

1 Introduction

The objective of this work has been to devise predictive algebraic expressions with a theoretically based structure for fully developed turbulent flow and convection in channels. These predictive expressions for flow are based on numerically values obtained by integration of differential models that are exact except for a semi-empirical correlative expression for the local time-averaged turbulent shear stress. The predictive expressions for convection introduce some idealizations but no further empiricism.

2 Development of the New Model

The starting point is the differential balance obtained by twice integrating the time-averaged Navier-Stokes equations for fully developed flow with invariant physical properties in a round tube,namely,

$$\tau_w\left(1-\frac{y}{a}\right)=\mu\frac{du}{dy}-\rho\overline{u'v'} \tag{1}$$

Rather than introducing the eddy viscosity or the mixing-length to represent $-\rho\overline{u'v'}$, and then devise correlative or predictive expressions for these heuristic quantities, (1) was simply re-expressed as

$$\left(1-\frac{y^+}{a^+}\right)\left[1-\left(\overline{u'v'}\right)^{++}\right]=\frac{du^+}{dy^+} \tag{2}$$

where $\left(\overline{u'v'}\right)^{++}\equiv-\frac{\rho\overline{u'v'}}{\tau}$ is the fraction of the total local shear due to the turbulent fluctuations. A correlating equation for this latter quantity is

$$\left(\overline{u'v'}\right)^{++}=\left(\left[0.7\left(\frac{y^+}{10}\right)^3\right]^{8/7}+\left|\exp\left\{\frac{-1}{0.436y^+}\right\}-\frac{1}{0.436a^+}\left(1+\frac{6.95y^+}{a^+}\right)\right|^{-8/7}\right)^{-7/8} \tag{3}$$

The 3rd-power dependence on y^+ is from asymptotic expansion, and the coefficient of 0.0007 from direct numerical simulations. The exponential term corresponds to the semi-logarithmic dependence of the time-averaged velocity on y^+,which was derived by Millikan[1] by speculative dimensional analysis. The following term represents the wake. The coefficients 0.436 and 6.95 are based on recent experimental measurements of the time-averaged velocity by Zagarola [2]. The combining exponent of $-8/7$ is based on the measurements of the fluctuating velocities in a parallel-plate channel by Wei and Willmarth [5].

3 Integral Formulations

Re-expressing (3) in terms of $R = 1 - y^+$ and integrating formally yields

$$u^+ = \frac{a^+}{2}\int_{R^2}^{1}\left[1-\left(\overline{u'v'}\right)^{++}\right]dR^2 = \frac{a^+}{2}\left(1-R^2\right) - \frac{a^+}{2}\int_{R^2}^{1}\left(\overline{u'v'}\right)^{++}dR^2 \quad (4)$$

In turn, integrating u^+ over the cross-section of the tube yields

$$\left(\frac{2}{f}\right)^{1/2} = u_m^+ = \int_0^1 u dR^2 = \frac{a^+}{4}\int_0^1\left[1-\left(\overline{u'v'}\right)^{++}\right]dR^4$$
$$= \frac{a^+}{4} - \frac{a^+}{4}\int_0^1\left(\overline{u'v'}\right)^{++}dR^4 \quad (5)$$

These two integral provide valuable insight, but stepwise integration of the differential formulations is more efficient computationally.

3.1 Inter-relationships

Eliminating $\frac{du^+}{dy^+}$ between the differential energy balances in terms of $(\overline{u'v'})++$, the eddy viscosity, and the mixing length reveals that $\frac{\mu_1}{\mu} = \frac{\left(\overline{u'v'}\right)^{++}}{\left(1-\overline{u'v'}\right)^{++}}$ and $(\ell^+) = \frac{\left(\overline{u'v'}\right)^{++}}{\left(1-\frac{y}{a}\right)\left[1-\overline{u'v'}^{++}\right]^2}$. Despite its heuristic origin the eddy viscosity is seen to be a physically meaningful quantity whereas the mixing-length is proven, perhaps for the first time, to be singular at the axis for all conditions because $\left(\overline{u'v'}\right)^{++}$ is always finite.

3.2 Correlating Equation for the Fanning Friction Factor

The numerically computed values of u_m^+ are represented almost exactly, and the experimental values of Zagarola [4] within 0.5%, by the following expression:

$$\left(\frac{2}{f}\right)^{1/2} = u_m^+ = 3.2 - \frac{227}{a^+} + \left(\frac{50}{a^+}\right)^2 + \frac{1}{0.436}ln\{a^+\} \quad (6)$$

No additional empiricism is introduced by (6). The unfamiliar terms in $(a^+)^{-1}$and $(a^+)^{-2}$ are the first terms in an asymptotic expansion for the effect of the viscous boundary layer as represented by $u^+ = y^+$.

3.3 Other Geometries

Space does not permit a detailed description of the development of analogous formulations and results for other geometries, but briefly: (1)-(3) are directly applicable for a parallel-plate channel if $a$is replaced by b; (3) may be adapted separately for the inner and outer regions of annuli; but empirical expressions are required for the locations of the maximum in the velocity and the zero in the shear stress. Equation (5) with $a_2^+ - a_1^+$ substituted for a^+provides a good approximation for annuli of all aspect ratios.

3.4 Analogous Expressions for Convection

The differential energy balance for a uniformly heated round tube may be expressed as

$$\frac{(1+\gamma)\left(1-\frac{y^+}{a^+}\right)}{1+\frac{Pr}{Pr_t}\left(\frac{\left(\overline{u'v'}\right)^{++}}{1-\left(\overline{u'v'}\right)^{++}}\right)} = \frac{dT^+}{dy^+} \tag{7}$$

where $T^+ = \dfrac{k\left(\rho\tau_w\right)^{1/2}\left(T_w - T\right)}{\mu_j}$ and $1+\gamma = \dfrac{1}{R^2}\displaystyle\int_0^{R^2}\left(\frac{u^+}{u_m^+}\right)dR^2$

Formal integration of (7) results in

$$T^+ = \frac{a^+}{2}\int_{R^2}^{1}\frac{(1+\gamma)\,dR^2}{1+\frac{Pr}{Pr_t}\left(\frac{\left(\overline{u'v'}\right)^{++}}{1-\left(\overline{u'v'}\right)^{++}}\right)} = \frac{dT^+}{dy^+} \tag{8}$$

and

$$\frac{2a^+}{Nu} = T_m^+ = \frac{a^+}{4}\int_0^1\frac{(1+\gamma)^2\,dR^4}{1+\frac{Pr}{Pr_t}\left(\frac{\left(\overline{u'v'}\right)^{++}}{1-\left(\overline{u'v'}\right)^{++}}\right)} = \frac{dT^+}{dy^+} \tag{9}$$

Again, numerical integration of the differential formulations is more efficient computationally.

3.5 The Utility of Analogies

Because of the insight provided by (8), the classical analogy of Reichardt [3], after the correction of several errors, was recognized as equivalent to

$$\frac{1}{Nu} = \left(\frac{Pr_t}{Pr}\right)\frac{1}{Nu_1} + \left(1-\frac{Pr_t}{Pr}\right)\frac{1}{Nu_\infty} \tag{10}$$

with

$$Nu_1 \equiv Nu\{Pr_t = Pr\} = \frac{8}{\int_0^1 (1+\gamma)^2\left[1-\left(\overline{u'v'}\right)^{++}\right]dR^4} = \frac{Re\,f/2}{(1+\gamma)^2_{wmR^4}}$$

and

$$Nu_\infty \equiv Nu\{Pr \longrightarrow \infty\} = 0.07343\left(\frac{Pr}{Pr_t}\right)^{1/3}\left(\frac{f}{2}\right)^{1/2}$$

Equation (10) proved to provide a very good representation for the computed values of Nu for all geometries, all thermal boundary conditions, and all values of$Pr > 0.7$. However, a slightly modified analogy derived by Churchill [4], which results in

$$Nu = \left(\left(\frac{Pr_t}{Pr}\right)\left[\frac{1}{Nu_1} + \left(\left[\frac{Pr}{Pr_t}\right]^{2/3} - 1\right)\frac{1}{Nu_\infty^1}\right]\right)^{-1} \tag{11}$$

is effectively exact. An analogue of (11) for $Pr \leq Pr_t$ is

$$\frac{Nu_1 - Nu}{Nu_1 - Nu_0} = \left(1 + \frac{(Pr_t/Pr)^{\frac{1}{8}}\,(Nu_1 - Nu_0)\,Nu_\infty^1}{\left(\frac{Pr_t}{Pr} - 1\right)\left(Nu_\infty^1 - \frac{2}{3}Nu_1\right)}\right)^{-1} \tag{12}$$

where $Nu_0 \equiv Nu\{Pr = 0\} = \frac{1}{8}\left(\int (1+\gamma)^2\, dR^4\right)^{-1} = \frac{1}{8}\left((1+\gamma)^2_{mR^4}\right)^{-1}$

Equations (11) and (12) have been found to represent the essentially exact numerical solutions for Nu within 1% for all values of Pr,all values of Re in the fully turbulent regime of flow in circular, concentric annuli of all aspect ratios including the limiting cases of a round tube and a parallel-plate channel, and all thermal boundary conditions that lead to fully developed convection (see [5] and [6] for illustrations).

4 Summary

An algebraic correlating eqution for the local fraction of the total shear stress due to the turbulent fluctuations allows prediction of the time-averaged velocity distribution and mixed-mean velocity in all one-dimensional flows in channels almost exactly for all values of Re. By virtue of an analogy between momentum and energy transfer, results of comparable accuracy are obtained without further empiricism for fully developed turbulent convection for all values of Pr.

References

1. Millikan, C.B., Proc. 5th Int. Congr. Appl. Mech., Cambridge, Mass.(1938) 386-392.
2. Zagarola, M., Mean-Flow Scalingof Turbulent Pipe Flow, Ph.D. Thesis, Princeton University (1996).
3. Reichardt, H. Archiv ges. Waermrtechnik. 2 (1951)129-132.
4. Churchill, S.W., Thermal Sci. Eng.5 (1997) 13-30.
5. Churchill, S.W., Zajic, AIChE J 48 (2002) 927-940.
6. YU, Bo, Kawaguchi, Y., Ozoe, H, Churchill, S.W., Intern. J.Heat Mass Transfer, in press.

Langevin Models of Turbulence

B. Dubrulle[1], J-P. Laval[2], and S. Nazarenko[3]

[1] Groupe Instabilité et Turbulence, SPEC/DRECAM/DSM, CNRS URA 2464, CEA Saclay, F-91191 Gif sur Yvette bdubrulle@cea.fr
[2] Laboratoire de Mécanique de Lille, CNRS UMR 8107, Bld Paul Langevin, F-59655 Villeneuve d'Ascq Jean-Philippe.Laval@univ-lille1.fr
[3] Mathematics Institute, University of Warwick, Coventry CV4 7AL, UK snazar@maths.warwick.ac.uk

1 Introduction

In a turbulent flow, the number of degree of freedom N can be gigantic, scaling as the 9/4 power of the Reynolds number. In the atmosphere, this number may reach $N \sim 10^{16}$, devastating our hope to implement all scales of the climate system onto a computer. This juggling with numbers illustrates the well known challenge posed by turbulent flows: is there a way to simulate, or describe a turbulent flow, without taking into account all degrees of freedom? A similar question has been asked in the past by founders of statistical mechanics. Specifically, it has been the kind of challenge met by Boltzman and co-workers to describe the behavior of a gas made of billions of particles. Of course, in the case of turbulence, an additional difficulty arises because a turbulent flow is necessarily driven out of equilibrium by the energy input mechanisms. Therefore, none of the well-known recipes of classical statistical mechanics apply. Yet, we may learn something from our glorious ancestors by closer inspection of their protocol: in a gas, the number of particles is so huge that it is just hopeless trying to follow each of them individually. Whatever our power of measurements, there will remain individual particles which we will be unable to follow. Instead of starting an endless race towards finer and finer measurements, aimed at decreasing their corresponding number, why not accept this inherent ignorance, and replace it by something mimicking its action, and which will be easy to handle? This is precisely the reasoning followed by Langevin, upon modeling the Brownian motion by a simple Gaussian white noise. Such simple rule achieved many successes. Can it be simply translated to turbulent flows? This possibility is discussed in the present short review. Additional point of view about this may be found in the contribution by Friedrich in these proceedings.

2 Langevin Models of Turbulence

2.1 Framework

Consider a turbulent flow, with velocity field $v_i(x,t)$, and introduce an (arbitrary) filtering procedure so as to separate it into a large-scale field $U_i = \overline{v_i}$ and a small-scale component $u_i = v_i - U_i$. Such small-scale motion varies over time scale t, while large scale vary over time scale T. In any reasonable turbulent flow, the ratio of the typical time scale of the two components varies like a power of the scale ratio, as $t/T \sim (l/L)(U/u) \sim (l/L)^{2/3}$. Therefore, small scales vary much more rapidly than large scale. From the point of view of the largest scales, the small scales may then be regarded as a noise. Hence the idea to simply replace them by an a priori chosen noise, with well-defined properties. One classical way is though a generalized Langevin equation:

$$\dot{u}_i = A_{ij}u_j + \xi_i, \tag{1}$$

where A is a generalized friction operator, and ξ is a noise. In the sequel, we explore various models characterized by different value of A and ξ.

2.2 Obukov Model

The simplest model one can imagine is to take $A = 0$ and ξ as a Gaussian white noise, isotropic and homogeneous in space, with short time correlation:

$$< \xi_i(x,t)\xi_j(x',t) >= 2\Delta\delta_{ij}\delta(t-t'). \tag{2}$$

This model has been first introduced in 1959 by Obukhov. It leads to a number of interesting properties.

Richardson Law and Kolmogorov Spectrum

Consider for example a cloud of passive scalar particles, embedded in such a flow. After a time t, this cloud of particles will have evolved into a situation where its velocity distribution obeys a Gaussian statistics, with variance scaling like square root of time: $\delta u = \sqrt{< u^2 > - < u >^2} \sim t^{1/2}$. In parallel, the cloud of particles experienced a spread by a factor $r \sim \sqrt{< x^2 > - < x >^2} \sim t^{3/2}$. This last law is nothing but the famous Richardson law, an empirical law describing the dispersion of passive tracers in the atmosphere. Moreover, we may combine the two simple relation to obtain that $\delta u \sim r^{1/3}$, implying a velocity spectrum $E(k) \sim k^{-5/3}$, i.e. the Kolmogorov spectrum. We see that with virtually no effort, Obukhov model reproduces the two more robust experimental results obtained so far in turbulence!

Limitations

Richardson law and Kolmogorov spectra are representative of velocities which do not differ from the mean by a large amount. The actual range of validity of the Obukhov model arises when considering higher moments, involving rarer, but more violent events. Since velocities in this model are Gaussian, their moments obey a simple scaling relation $:< u^{2n} >\sim< u^2 >^n$, at variance with the intermittency observed in real turbulent flows. This simple hierarchy law disappears as soon as one allows for spatial or temporal correlation, as recently proved in the Kraichnan model of turbulence.

2.3 Kraichnan Model

The Obukhov model is frictionless in essence. The Kraichnan model can be viewed as the opposite limit, with a very large friction $A_{ij} = -\gamma\delta_{ij}$, $\gamma \ll 1$, and a noise with spatial correlation

$$< \xi_i(x,t)\xi_j(x',t) >= 2\Delta_{ij}(x,x')\delta(t-t'). \tag{3}$$

Due to the large friction, the inertial term in the Langevin equation becomes negligible and the velocity adiabatically adjusts to the noise as: $u_i \sim \gamma\xi_i$. The Kraichnan model is thus made of small-scale delta-correlated Gaussian white noise, with spatial correlation.

Intermittency and Conservation Laws

Contrarily to Obukhov model, Kraichnan model leads to intermittency for the high order moments. The physical reasons have been recently reviewed in [1]. They are rooted in the spatial correlation, which induce a memory effect onto lagragian trajectories, and lead to the apparition of conservation laws within sets of Lagrangian particles. Since the moment of order $2n$ is associated with conservation laws of sets of $2n$ particles, and since conservation laws of sets of particles of different sizes are not simply related, this induces a breaking of the hierarchical structure of the moments.

Turbulent Transport

Another less well known property of Kraichnan model concerns turbulent transport. Suppose we focus on the evolution of the vorticity in such a model. In classical turbulence, the vorticity obeys the equation

$$\partial_t\Omega_i = -v_k\partial_k\Omega_i + \Omega_k\partial_k v_i + \nu\partial_k\partial_k\Omega_i, \tag{4}$$

where ν is the molecular viscosity, and v is the sum of the large scale component U and the (small-scale) noise. Because of the presence of noise, eq. (4) admits stochastic solution, whose dynamic can be fully specified by the probability distribution function. Ignoring the viscosity and using standard techniques [2], one can derive the evolution equation for $P(\Omega, x, t)$, the probability of having the field Ω at point x and time t:

$$\begin{aligned}\partial_t P = &-U_k\partial_k P - (\partial_k U_i)\partial_{\Omega_i}[\Omega_k P] + \partial_k[\beta_{kl}\partial_l P] \\ &+ 2\partial_{\Omega_i}[\Omega_k \alpha_{lik}\partial_l P] \\ &+ \mu_{ijkl}\partial_{\Omega_i}[\Omega_j \partial_{\Omega_k}(\Omega_l P)]\end{aligned} \tag{5}$$

For simplicity, we assumed homogeneity of the fluctuations and we introduced the following turbulent tensors:

$$\begin{aligned}\beta_{kl} &= \langle u_k u_l\rangle \\ \alpha_{ijk} &= \langle u_i \partial_k u_j\rangle \\ \mu_{ijkl} &= \langle \partial_j u^i \partial_l u^k\rangle\end{aligned} \tag{6}$$

Due to incompressibility, the following relations hold: $\alpha_{kii} = \mu_{iikl} = \mu_{ijkk} = 0$.

To illuminate the signification of this complicated equation, let us consider the first moment of eq. (5), obtained by multiplication with Ω_i and integration:

$$\begin{aligned}\partial_t\langle\Omega_i\rangle = &-U_k\partial_k\langle\Omega^i\rangle + (\partial_k U_i)\langle\Omega_k\rangle - 2\alpha_{kil}\partial_k\langle\Omega_l\rangle \\ &+ \beta_{kl}\partial_k\partial_l\langle\Omega_i\rangle.\end{aligned} \tag{7}$$

In addition to the standard vorticity advection and stretching by the large scale, one recognize two additional effect: one proportional to α, resulting in large-scale vorticity generation through the AKA instability [3]; one proportional to β, akin to a turbulent viscosity. Within the Kraichnan model, one therefore naturally recovers the well-known formulation of turbulent transport, without resorting to scale separation [4]. In this very simple model, where the viscosity has been ignored, one can show that the tensor β is always positive: the turbulent viscosity always enhances turbulent transport. In actual viscid flows, the turbulent viscosity tensor is actually fourth order, and can be negative [4].

Limitation

This digression about turbulent transport shows that the way we prescribe velocity correlation in Kraichnan model somehow determines the turbulent transport properties of the flow. It is a kind of adjustable parameter. In that respect, it would be nice to devise a model devoid of this freedom of choice, by ensuring for example that the turbulent transport somehow adjusts itself to the way energy is injected and dissipated, as in real turbulence. In the sequel, we present a model where the noise is dynamically computed at each time scale, thereby removing the arbitrariness of the Langevin model.

2.4 Stochastic RDT Model

Description

Our method is based on the observation that small scales are mostly slaved to large scale via linear processes akin to rapid distortion. This observation is substantiated by various numerical simulations and is linked with the prominence of non-local interactions at small scale [5]. Specifically, let us decompose our small-scale velocity field into wave packets, via a localized Fourier transform:

$$\hat{u_i}(x,k) = \int h(x-x')e^{ik(x-x')}u_i(x')dx',$$

where h is a filtering function, which rapidly decays at infinity. Using incompressibility and non-locality of interaction, one can derive the following equation of motion for the wave-packet [5, 6]:

$$\begin{aligned} \dot{x}_i &= U_i, \\ \dot{k}_i &= -k_j \partial_j U_i, \\ \dot{\hat{u}}_i &= -\nu_t k^2 \hat{u}_i + \hat{u}_j \partial_j \left(2\frac{k_i k_m}{k^2} U_m - U_m \right) + \hat{\xi}_i. \end{aligned} \quad (8)$$

Here ν_t is a turbulent viscosity describing the local interactions between small-scales, and ξ is a forcing stemming from the energy cascade. Its expression only involves large-scale non-linearities (in fact the aliasing) via $\xi_i = \partial_j \left(U_j U_i - \overline{U_i U_j} \right)$. By eq. (8) the wave-packet is transported by the large-scale flow, its local wavenumber is distorted by the large-scale velocity gradients, and its amplitude is modified through the action of local and non-local interactions. The equation describing its amplitude evolution is a generalized Langevin equation, with friction generated by turbulent viscosity and with both multiplicative and additive noise stemming from interaction with large scale. Because these two noises are of same origin, they are correlated. One can show that this correlation is responsible for a skewness in the probability distribution of the small scale [5]. Note also that in some sense, our Langevin model can be viewed as a generalization of Rapid Distortion Theory equation, with inclusion of turbulent viscosity and stochastic forcing. No wonder, interesting analytical properties will be available in precisely the same case where Rapid Distortion Theory is the most useful, namely rotating, or stratified shear flows (see below).

Equation (8) shows that our model is specified by the knowledge of ξ and U_i. The latter can be shown by mere filtering to obey the equation:

$$\partial_t U_i + U_j \partial_j U_i = -\partial_i \bar{p} + \nu \Delta U_i + \bar{f}_i, \quad (9)$$

where f_i describes the backreaction of small scales onto large scales, and is obtained through summation over wavepackets:

$$f_i(x) = \int dk \partial_j \left(\overline{U_i(x)\hat{u}_j(x,k)} + \overline{(\hat{u}_i(x,k)U_j(x)} + \overline{(\hat{u}_i(x,k)\hat{u}_j(x,-k)} \right).$$

The set of eqs. (8) and (9) is a strongly nonlinear system of coupled equations, which defines our turbulence langevin model. In this method, the noises can be dynamically computed at each time step by integration of the large -scale equation. In the sequel, we present two application of this: one in which the system is simplified by *prescribing* one of the forcing (namely ξ). This allows the computation of general scaling laws for turbulent transport in various systems. One in which ξ is numerically computed using the large scale equation. This allows for fast numerical simulations.

Turbulent Transport

We consider here a simplified version of our Langevin model where the function ξ is not computed, but prescribed as a Gaussian delta-correlated white noise. The advantage of this simplification is that it allows for analytical computation in special situation, where the geometry of the system is so simple that it allows for explicit solution of the homogeneous (unforced) small-scale equations. In some sense, our model with prescribed Gaussian model for ξ can be viewed as a generalization of Kraichnan model of turbulence. One can then expect this model to provide "reasonable shape" for turbulent transport (see Section 2.3), with "free" parameter induced by the prescribed correlation function for ξ. Working out the details, we found out that indeed, our model is able to provide the scaling of the turbulent transport, as a function of control parameters, up to a numerical prefactor, controlled by the intensity of the correlation of the forcing ξ (the "free parameter").

Heat Transport in Convection

When a horizontal layer of fluid is heated from below, a heat exchange from the top to the bottom occurs. The transport of heat depends on the interplay between the thermal, viscous and integral scales of turbulence, and thus, on both the Prandtl number and the Reynolds numbers. Our model can be used to predict both the structure and the scaling laws in thermal convection [7]. In the boundary layer, the velocity profile is logarithmic and the temperature decays like the inverse of the distance to the wall. This has important impact onto the heat transport. At low Reynolds numbers, when most of the dissipation comes from the mean flow, we recover power classical scaling regimes of the Nusselt versus Rayleigh number, with exponent 1/3 or 1/4. At larger Reynolds number, velocity and temperature fluctuations become non-negligible in the dissipation. In these regimes, there is no exact power law dependence the Nusselt versus Rayleigh or Prandtl. Instead, we obtain logarithmic corrections to the classical soft (exponent 1/3) or ultra-hard (exponent 1/2) regimes, in a way consistent with the most accurate experimental measurements available nowadays. Example is given in the figure 1, showing the comparison between the data of the Castaing group in Helium, versus the theoretical predictions (lines).

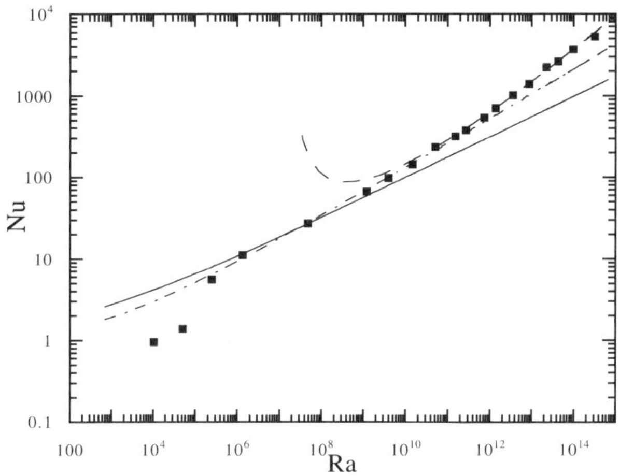

Fig. 1. Illustration of the three scaling regimes found in convection in Helium for Nusselt vs Rayleigh. The symbols are experimental measurements by [8]. The lines are theoretical prediction by [7] using an analytical model of turbulent convection. "Soft" turbulence regime(mean flow dominated): power law $Nu \sim Ra^{1/4}$ (full line); "Hard" turbulence regime: (velocity fluctuation dominated) $Nu \sim Ra^{1/3}/(\ln(Ra))^{2/3}$ (dotted line); "Ultra-hard" turbulent regime: (temperature fluctuations dominated) $Nu \sim Ra^{1/2}/(\ln(Ra))^{3/2}$ (dashed line)

The theory has also been extended to describe turbulent thermal convection at large Prandtl number [9]. Two regimes arise, depending on the Reynolds number Re. At low Reynolds number, $NuPr^{-1/2}$ and Re are a function of Ra $Pr^{-3/2}$. At large Reynolds number $NuPr^{1/3}$ and $RePr$ are function only of $RaPr^{2/3}$ (within logarithmic corrections). In practice, since Nu is always close to $Ra^{1/3}$, this corresponds to a much weaker dependence of the heat transfer in the Prandtl number at low Reynolds number than at large Reynolds number. This difference may solve an existing controversy between measurements in SF6 (large Re) and in alcohol/water (lower Re). These regimes may be linked with a possible global bifurcation in the turbulent mean flow. A scaling theory can be used to describe these two regimes through a single universal function. This function presents a bimodal character for intermediate range of Reynolds number. This bimodality can be explained in term of two dissipation regimes, one in which fluctuation dominate, and one in which mean flow dominates. Altogether, our results provide a six parameters fit of the curve $Nu(Ra, Pr)$ which may be used to describe all measurements at $Pr > 0.7$.

Momentum Transport in Rotating Shear Flow

At sufficiently large Reynolds number, the fluid between co-rotating coaxial cylinders becomes turbulent, and a significant momentum transport occurs between the two cylinders. In the case with rotating inner cylinder and resting outer one (the so-called Taylor-Couette flow), detailed measurements show that the torque applied at cylinders by the turbulent flow is a function of the Reynolds number R. Within the Langevin model, one can work out an analogy between the problem of momentum transport and heat transport in turbulent convection, to compute the torque in Taylor-Couette configuration, as a function of the Reynolds number [10]. At low Reynolds numbers, when most of the dissipation comes from the mean flow, we predict that the non-dimensional torque $G = T/\rho\nu^2 L$, where L is the cylinder length, scales with Reynolds number R and ratio of inner cylinder to outer cylinder radius $\eta = r_i/r_o$, $G = 1.46\eta^{3/2}(1-\eta)^{-7/4}R^{3/2}$. At larger Reynolds number, velocity fluctuations become non-negligible in the dissipation. In these regimes, there is no exact power law dependence the torque versus Reynolds. Instead, we obtain logarithmic corrections to the classical ultra-hard (exponent 2) regimes:

$$G = 0.50\frac{\eta^2}{(1-\eta)^{3/2}}\frac{R^2}{\ln[\eta^2(1-\eta)R^2/10^4]^{3/2}}.$$

These predictions are found to be in excellent agreement with available experimental data (see figure 2).

Fast Numerical Simulations

We consider now the case where ξ is not prescribed, but dynamically computed using the large scale equation. In that case, there is no free parameter in the problem, except for the magnitude of the turbulent viscosity. By comparison with direct numerical simulation, we found however [6] that in isotropic case, the magnitude of this turbulent viscosity can be prescribed using the DSTA model of Kraichnan

$$\nu_t(k) = C_t\left(\frac{k}{k_i}\right)^{-4/3}\sqrt{\frac{E(k_i)}{k_i}}, \tag{10}$$

where E is the energy spectrum, k_i is a wavenumber in the inertial range and C_t is a constant depending on a parameter β characterizing the degree of non-locality of the interaction. For the contribution at k of all modes with wavenumber greater than βk, it yields:

$$C_t(\beta) = \sqrt{7/60}\,\beta^{-2/3} = 0.3416\,\beta^{-2/3}, \tag{11}$$

with β depending on the ratio of the largest wave-number of the (resolved) simulation onto the cut-of wavenumber as $\beta = k_{max}/k_c$.

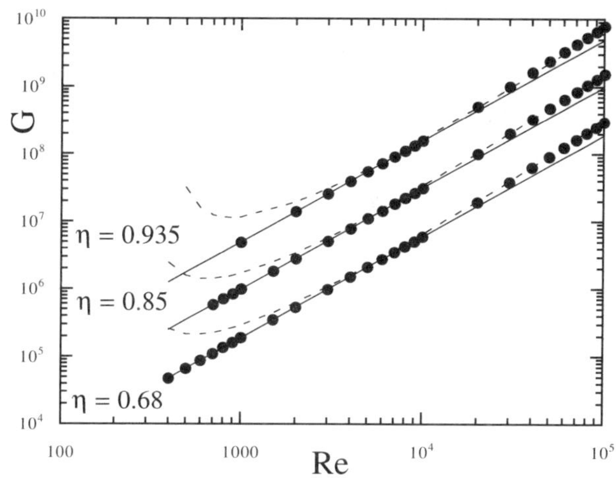

Fig. 2. Torque vs Reynolds in Taylor-Couette experiments for different gap widths $\eta = 0.68$, $\eta = 0.85$ and $\eta = 0.935$. The symbols are the data of [11]. The lines are the theoretical formula obtained in the soft and ultra-hard turbulence regimes and computed using the analogy with convection. Soft turbulence (full line); ultra-hard turbulence (dotted line). There is no adjustable parameter in this comparison, all the constants being fixed either by the analogy with convection, or by the comparison with the data of [12].

With this prescription, we may then see our Langevin model as a parameter-free model of turbulence. Its formulation is rather complex, but its advantage lies in the possibility to use a semi-Lagrangian scheme of integration for the small scale, thereby allowing for very large time steps. As a result, we obtain a fast numerical simulation, with all scales being resolved, but with an integration time smaller by a factor 10 to 1000 with respect to traditional DNS [13]! An example is provided on Figure 3 in the case of 2D turbulence.

3 Towards a LES Langevin Model?

In this short review, we hope to have convinced you of the interest of Langevin models of turbulence. However, we did not yet fully achieve the goal we fixed in the introduction: our model still retains infinitely many degrees of freedom, symbolized by unrestrained number of wave-packets we use. In some situations, this feature is more than desirable: a peasant working on his crop is seldom interested by the weather forecast at the level of his country, and would like to know the hail forecast at the level of his field! However, in most applications, one does not need such a wealth of details, and one would prefer a Langevin model with very few degrees of freedom. Such a "large eddy

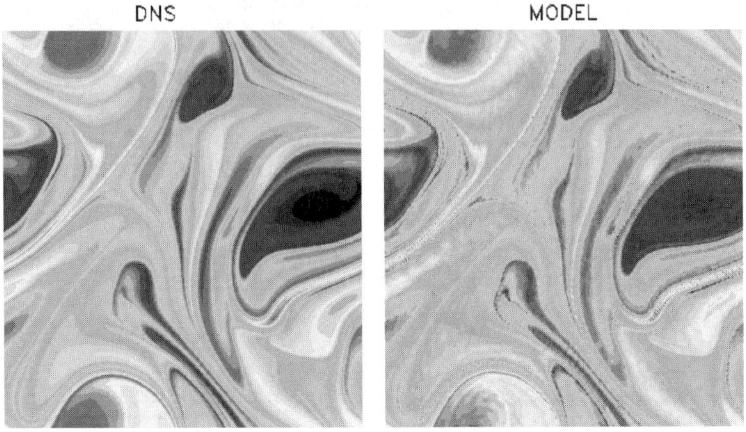

Fig. 3. Fast 2D numerical simulation. Left panel: vertical vorticity $\omega_z = \nabla\times\mathbf{u}\cdot e_z$ computed using standrad spectral method. This simulation required 3 days to be completed on our workstation. Right panel: same field, computed using our Langevin method. This was obtained in only 30 minutes, on the same work station.

Langevin" model remains to be built. We are currently working in that direction.

References

1. G. Falkovich, K. Gawedzki, and M. Vergassola. *Rev. Mod. Phys.*, 73:913–975, 2001.
2. N. Leprovost and B. Dubrulle. *Phys. Rev. Letter*, submitted, 2004.
3. U. Frisch, Z. S. She, and P. L. Sulem. *Physica D*, 28:382–392, 1987.
4. B. Dubrulle and U. Frisch. *Phys. Rev. A*, 43:5355–5364, 1991.
5. J.-P. Laval, B. Dubrulle, and S. Nazarenko. *Phys Fluids*, 13:1995–2012, 2001.
6. J.-P. Laval, B. Dubrulle, and J. C. McWilliams. *Phys. Fluids*, 15(5):1327–1339, 2003.
7. B. Dubrulle. *European Phys. Journal B*, 21:295–304, 2001.
8. X. Chavanne, F. Chilla, B. Castaing, B. Hébral, B. Chabaud, and J. Chaussy. *Phys. Rev. Lett.*, 79:3648–3651, 1997.
9. B. Dubrulle. *European Phys. Journal B*, 28:361–367, 2002.
10. B. Dubrulle and F. Hersant. *Eur. Phys. J. B*, 26(3):379–386, 2002.
11. F. Wendt. *Ingenieur-Archiv.*, 4:577–595, 1933.
12. G.S. Lewis and H.L. Swinney. *Phys Rev. E*, 59(004):5457–5467, 1999.
13. J.-P. Laval, B. Dubrulle, and S. Nazarenko. *submitted to J. Comp. Phys.*, 2000.

Renormalized Perturbation Theory for Lagrangian Turbulence

Rudolf Friedrich and Rafaela C. Hillerbrand

Institut für Theoretische Physik Westfälische Wilhelms-Universität
fiddir@uni-muenster.de, rafaelahillerbrand@uni-muenster.de

1 Outline

Recent experimental results on particle tracking have revealed the existence of intermittent behaviour of turbulent Lagrangian velocity statistics [1, 2]. In [3] attempts have been reported to derive the single particle statistics from the Navier-Stokes equation. This approach is based on a closure of the hierachy of evolution equations for the *p*robability *d*istribution *f*unctions (pdfs) of Lagrangian particles. In the present article we point out the necessity of performing a renormalized closure. This renormalized perturbation expansion leads to solutions for the pdfs of the velocity increments that belong to the class of *c*ontinuous *t*ime *r*andom *w*alks (ctrws).

2 Lagrangian Description of Turbulence

One of the many advantages of the Lagrangian approach to turbulence is that due to the Newtonian form of the equations of motion any modelling assumption for the acceleration field $\mathbf{A}$ will preserve the Galilean invariance of the underlying hydrodynamic equations. The Lagrangian point of view considers fluid motion as a transformation $\mathbf{X}(t;\mathbf{w},\mathbf{y})$, mapping the initial position $\mathbf{X}(t=0;\mathbf{w},\mathbf{y})=\mathbf{y}$ of a fictious tracer particle with initial velocity $\mathbf{w}$ onto its position $\mathbf{X}$ at later times $t>0$. The equations of motion then take the following form:

$$\frac{\mathrm{d}\mathbf{X}(t;\mathbf{w},\mathbf{y})}{\mathrm{d}t}=\mathbf{U}(t;\mathbf{w},\mathbf{y})\,,\quad \frac{\mathrm{d}\mathbf{U}(t;\mathbf{w},\mathbf{y})}{\mathrm{d}t}=\mathbf{A}(t;\mathbf{w},\mathbf{y}),\tag{1}$$

where the particle's acceleration $\mathbf{A}$ can be expressed in terms of the Eulerian pressure and velocity field evaluated at the location of the particle [4]. Via the acceleration the pressure term introduces long-range interactions between distant particles.

The main quantities of interest in the statistical theory of Lagrangian turbulence are the pdfs of the fluid particles [5]. For a single particle whose evolution is given by equation (1) the pdf is defined as

$$f(\mathbf{u},\mathbf{x},t) \equiv \langle \delta(\mathbf{u}-\mathbf{U}(t;\mathbf{w},\mathbf{y}))\delta(\mathbf{x}-\mathbf{X}(t;\mathbf{w},\mathbf{y}))\rangle\,, \tag{2}$$

where the brackets $\langle\cdot\rangle$ denote a suitable average. All pdfs shall be normalized to 1. The Eulerian statistics are obtained by integrating over all initial positions $\mathbf{y}$. For incompressible flows the Jacobian det $\frac{\partial X_i}{\partial y_j}$ of the transformation $\mathbf{X}(t;\mathbf{w},\mathbf{y})$ equals 1.

3 Evolution of Lagrangian pdfs

3.1 Hierachy of Evolution Equations

The starting point is an infinite set of evolution equations for the N particle joint position-velocity pdf [4], which can be derived from the Navier-Stokes equations in a similar way as the one for the Eulerian statistics considered by Lundgren [6]. The Lagrangian pdf for a single particle as defined in equation (2) obeys the following equation:

$$\left(\frac{\partial}{\partial t}+\mathbf{u}\cdot\nabla_{\mathbf{x}}\right)f(\mathbf{u},\mathbf{x},t) = -\nabla_{\mathbf{u}}\cdot\int \mathrm{d}^3\mathbf{v}\mathrm{d}^3\mathbf{r}\mathbf{A}(\mathbf{x}-\mathbf{r},\mathbf{v})f(\mathbf{v},\mathbf{r},t;\mathbf{u},\mathbf{x},t). \tag{3}$$

Here $\mathbf{x}$, $\mathbf{u}$ again denote location and velocity of a Lagrangian particle while $\mathbf{v}$ and $\mathbf{r}$ are Eulerian fields at time t. The acceleration $\mathbf{A}(\mathbf{x}-\mathbf{r},\mathbf{v})$ is related to the pressure and the dissipation term in the Navier-Stokes equation as

$$\mathbf{A}(\mathbf{x}-\mathbf{r},\mathbf{v}) \equiv -\nabla\frac{\partial^2}{\partial x_i x_j}\frac{1}{4\pi|\mathbf{x}-\mathbf{r}|}v_i v_j + \nu\delta\left(\mathbf{x}-\mathbf{r}\right)\triangle_{\mathbf{r}}\mathbf{v}. \tag{4}$$

This evolution equation for the one-particle pdf entails higher order statistics via the mixed Eulerian-Lagragian pdf $f(\mathbf{v},\mathbf{r},t;\mathbf{u},\mathbf{x},t') \equiv \langle\delta(\mathbf{v}-\mathbf{V}(\mathbf{r},t),\,\mathbf{u}-\mathbf{U}(t',\mathbf{w},\mathbf{y}),\mathbf{x}-\mathbf{X}(t',\mathbf{w},\mathbf{y}))\rangle$. The evolution of this mixed pdf itself depends on the higher order distribution function $f(\mathbf{v}',\mathbf{r}',t';\mathbf{v},\mathbf{r},t;\mathbf{u},\mathbf{x},t')$, where $\mathbf{v}$ and $\mathbf{v}'$ are Eulerian fields at the points $(\mathbf{r},t)$ and $(\mathbf{r}',t')$, respectively:

$$\left(\frac{\partial}{\partial t'}+\mathbf{u}\cdot\nabla_{\mathbf{x}}\right)f(\mathbf{v},\mathbf{r},t;\mathbf{u},\mathbf{x},t') = -\nabla_{\mathbf{u}}\cdot\int \mathrm{d}^3\mathbf{v}'\mathrm{d}^3\mathbf{r}'\mathbf{A}(\mathbf{x}-\mathbf{r}',\mathbf{v}')\, f(\mathbf{v}',\mathbf{r}',t';\mathbf{v},\mathbf{r},t;\mathbf{u},\mathbf{x},t'). \tag{5}$$

As in the case of the one-particle distribution in equation (3) no closed equation can be obtained for the higher order statistics.

3.2 Renormalized Mean-Field Approach

The hierachy indicated by equations (3) and (5) can be closed by making assumptions about the way in which distribution functions factorise. The simplest closure assumption neglects three-point correlations, i.e.

$$f(\mathbf{v}',\mathbf{r}',t';\mathbf{v},\mathbf{r},t;\mathbf{u},\mathbf{x},t') \approx f(\mathbf{v}',\mathbf{r}',t';\mathbf{v},\mathbf{r},t)f(\mathbf{u},\mathbf{x},t'). \quad (6)$$

Inserting this approximation in equation (5), the right-hand side vanishes for stationary turbulence when integrating over $\mathbf{r}$ and $\mathbf{v}$. Therefore such a mean-field approach can only be consistently made when at the same time the "free propagator" $G_0(\mathbf{x},\mathbf{u},t;\bar{\mathbf{x}},\bar{\mathbf{u}},\bar{t}) = \left(\frac{\partial}{\partial t} + \mathbf{u}\cdot\nabla_\mathbf{x}\right)^{-1}\delta(\mathbf{u}-\bar{\mathbf{u}},\mathbf{x}-\bar{\mathbf{x}},t-\bar{t})$ is replaced by the renormalized propagator $G(\mathbf{x},\mathbf{u},t';\bar{\mathbf{x}},\bar{\mathbf{u}},\bar{t}')$ given via the line-renormalized version of equation (5):

$$\left(\frac{\partial}{\partial t'} + \mathbf{u}\cdot\nabla_\mathbf{x}\right) f(\mathbf{v},\mathbf{r},t;\mathbf{u},\mathbf{x},t') - \int \mathrm{d}^3\bar{\mathbf{x}}\mathrm{d}^3\bar{\mathbf{u}}\mathrm{d}\bar{t}'\,\Sigma(\mathbf{x},\mathbf{u},t';\bar{\mathbf{x}},\bar{\mathbf{u}},\bar{t}')\cdot$$

$$f(\mathbf{v},\mathbf{r},t;\bar{\mathbf{x}},\bar{\mathbf{u}},\bar{t}') = -\nabla_\mathbf{u}\cdot\int d^3\mathbf{v}'d^3\mathbf{r}'\mathbf{A}(\mathbf{x}-\mathbf{r}',\mathbf{v}')f(\mathbf{v}',\mathbf{r}',t';\mathbf{v},\mathbf{r},t)f(\mathbf{u},\mathbf{x},t'). (7)$$

Integrating over the Eulerian field variables $\mathbf{v}$ and $\mathbf{r}$ leads to the line-renormalized transport equation for the single-particle pdf $f(\mathbf{u},\mathbf{x},t)$:

$$\left(\frac{\partial}{\partial t} + \mathbf{u}\cdot\nabla_\mathbf{x}\right) f(\mathbf{u},\mathbf{x},t) = \int \mathrm{d}^3\bar{\mathbf{x}}\mathrm{d}^3\bar{\mathbf{y}}\mathrm{d}\bar{t}\,\Sigma(\mathbf{x},\mathbf{u},t;\bar{\mathbf{x}},\bar{\mathbf{u}},\bar{t})\,f(\bar{\mathbf{u}},\bar{\mathbf{x}},\bar{t}). \quad (8)$$

This allows the identification of Σ with the self-energy of the system. Inserting the solution of equation (7) into the original evolution equation of the single-particle pdf (3) the contributions of the boundary and the homogeneous term vanish when we assume the initial conditions to have the form $f(\mathbf{v},\mathbf{r},t;\mathbf{u},\mathbf{x},t'=0) = f(\mathbf{v},\mathbf{r},t)f(\mathbf{u},\mathbf{x},t'=0)$. Then comparing equation (3) to the renormalized equation (8) leads to an expression of the self-energy Σ in terms of the renormlized propagator G:

$$\Sigma(\mathbf{x},\mathbf{u},t;\bar{\mathbf{x}},\bar{\mathbf{u}},\bar{t}) = \nabla_\mathbf{u}\langle\mathbf{A}(\mathbf{x},t)\mathbf{A}(\bar{\mathbf{x}},\bar{t})\rangle G(\mathbf{x},\mathbf{u},t;\bar{\mathbf{x}},\bar{\mathbf{u}},\bar{t})\nabla_{\bar{\mathbf{u}}}. \quad (9)$$

The expression for Σ given in [3] contains the unrenormalized propagator G_0 instead of the renormalized one.

If the acceleration-acceleration correlations $\langle\mathbf{A}(\mathbf{x},t)\mathbf{A}(\bar{\mathbf{x}},\bar{t})\rangle$ are given, the equation is closed since G is just the single particle pdf of a particle starting at time $\bar{t}$ at $\bar{\mathbf{x}}$. As a result, we see that a renormalized closure of the hierarchy of evolution equations for Lagrangian particles yields nonlinear generalized Fokker-Planck equations.

4 Continuous Time Random Walks (CTRWs)

Starting from the line-renormalized evolution equation for the single particle pdf, equation (8), an analogy to ctrws can be drawn. In fact, the evolution equation for the pdf of the velocity increment V, $h(V,t)$, takes the same form as given in [3] and the same reasoning can be taken over to the renormalized equation. We point out here that the underlying stochastic process is a ctrw. It is quite interesting to look at the solutions of this equation, which are in close relation to the pdfs obtained in [7]:

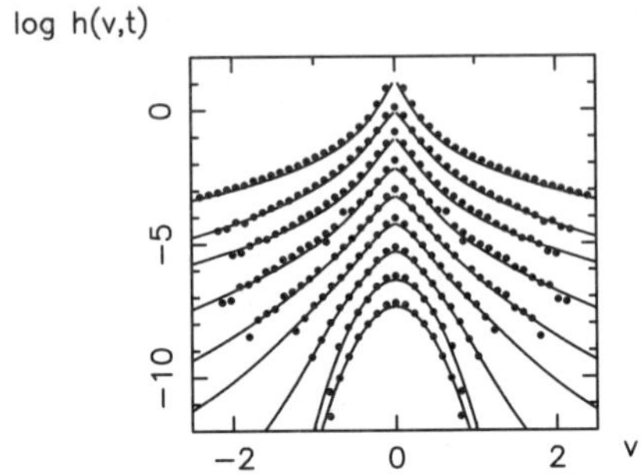

Fig. 1. pdfs h for the velocity increment V: — theoretical values obtained from equation (10), • experimental data from [2].

$$h(V,t) = \int \mathrm{d}s \frac{1}{\alpha} \frac{t}{s^{1+1/\alpha}} l_\alpha \left(\frac{t}{s^{1/\alpha}} \right) e^{-V^{2\alpha}/(4Q_0\alpha^2 s)} \frac{1}{(4Q_0\alpha^2 s)^{3/(2\alpha)}} , \qquad (10)$$

where $l_\alpha(\xi)$ denotes the one-sided Lévy distribution of order α (see [3]). Furthermore it was shown in [3], how these solutions relate to the class of ctrws as investigated in [8]. The pdfs for the velocity increments obtained from experiment can be reproduced well by fixing the only free parameter in equation (10), $0 < \alpha < 1$ (see fig.1).

Summarizing, we have outlined a renormalized closure approach for the derivation of Lagrangian particle pdfs. It is hoped that such an approach will lead to an assessement of the stochastic properties of Lagrangian particles in turbulence.

References

1. La Porta A, Voth G, Crawford AM, Alexander J, Bodenschatz E (2001) Nature 409:1017
2. Mordant N, Metz P, Michel O, Pinton J-F (2001) Phys Rev Lett 87:214501
3. Friedrich R (2003) Phys Rev Lett 90: 084501
4. Friedrich R (2003) arXiv:physics/0207015
5. Pope SB (2001) Turbulent Flows. University Press, Cambridge
6. Lundgren TS (1967) Phys Fluids 10:969
7. Chanal O, Castaing B, Hébral B (2000) Eur Phys J B 17:301
8. Barkai E (2001) Phys Rev E 63:046118

Modelling of the Pressure-Strain- and Diffusion-Term in Rotating Flows

S. Guenther and M. Oberlack

Hydromechanics and Hydraulics Group, Technische Universität Darmstadt, Petersenstraße 13, 64287 Darmstadt, Germany, guenther@hyhy.tu-darmstadt.de oberlack@hyhy.tu-darmstadt.de

1 Introduction

We reconsider the problem of shear free turbulent diffusion in a rotating frame, rotating about x_1. Therefore turbulence is generated at a vibrating grid in the $x_2 - x_3$ plane and diffuses away from the grid in x_1-direction. The flow geometry is given in figure 1. An important property of this flow is that

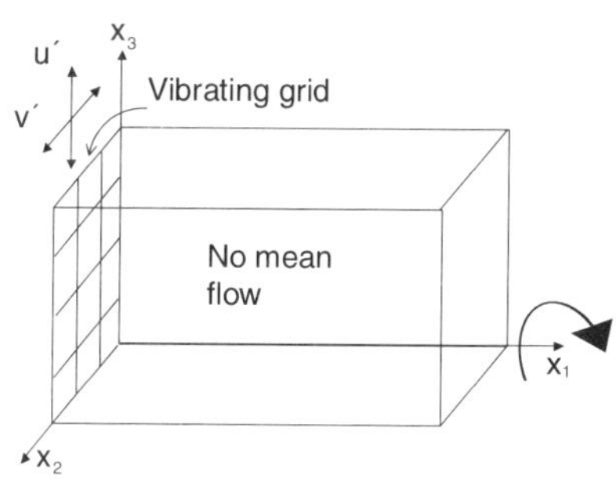

Fig. 1. Flow geometry

there is no mean flow-velocity. With the help of Lie-group methods Reynolds-stress transport models can be analyzed for this kind of flow in a rotating frame. From the analysis it can be found that the turbulent diffusion only influences a finite domain. Insertion of this solution in the model equations shows that even fully nonlinear Reynolds-stress transport models (nonlinear in the Reynolds-stresses for the pressure-strain model) are insensitive to rotation for this type of flow. To improve this serious shortcoming of these models, a new model equation which is fully nonlinear in the mean-velocity gradient is needed. There appears to be one one-point model which may account for system rotation for this type of flow. It is the second-moment closure model by

Sjögren and Johansson [3] which is nonlinear in the mean-velocity gradient. Therefore new model equations for the pressure-strain-correlation and the diffusion term have been developed along the lines of this model.

2 New Model of the Pressure-Strain-Term

In modelling the pressure-strain-correlation it is assumed, that this term depends on the mean shear- (S_{ij}), mean rotation- (W_{ij}) and the Reynolds-stress anisotropy-tensor (b_{ij}). It can be further split up into three parts which are the classical return- and rapid-term and a new term which will be called nonlinear-scrambling-term (nls).

To model the return- and rapid-term the Cayley-Hamilton-Theorem is applied leading to an approach which is nonlinear in b_{ij}. With the condition of no mean flow velocity the model consists of five tensor invariants since S_{ij} is zero and W_{ij} contains only Ω_i. The coefficients are functions of the scalar invariants. Introducing the variables

$$\phi_{ij}^{nls^\star} = \frac{\phi_{ij}^{nls}}{\epsilon} \quad , \qquad S_{ij}^\star = \frac{k}{\epsilon} S_{ij} \quad , \quad W_{ij}^\star = \frac{k}{\epsilon} W_{ij} \quad , \qquad b_{ij}^\star = b_{ij} \tag{1}$$

the non-liner-scrambling-term can be written in dimensionless form:

$$\begin{aligned} \phi_{ij}^{nls^\star} &= \beta_1 \left(W_{ij}^{\star 2} - \frac{1}{3} W_{ll}^{\star 2} \delta_{ij} \right) + \beta_2 \left(W_{ik}^{\star 2} b_{kj}^\star - \frac{1}{3} W_{lk}^{\star 2} b_{kl}^\star \delta_{ij} \right) \\ &+ \beta_3 \left(W_{ik}^{\star 2} b_{kj}^{\star 2} - \frac{1}{3} W_{lk}^{\star 2} b_{kl}^{\star 2} \delta_{ij} \right) + \beta_4 \left(W_{ik}^{\star 2} b_{kl}^\star W_{lj}^\star - \frac{1}{3} W_{hk}^{\star 2} b_{kl}^\star W_{lh}^\star \delta_{ij} \right) \\ &+ \beta_5 \left(W_{ik}^{\star 2} b_{kl}^{\star 2} W_{lj}^\star - \frac{1}{3} W_{hk}^{\star 2} b_{kl}^{\star 2} W_{lh}^\star \delta_{ij} \right) , \end{aligned} \tag{2}$$

with

$$\begin{aligned} \beta_i = f\,[\; & tr(b_{ij}^\star), tr(b_{ij}^{\star 2}), tr(b_{ij}^{\star 3}), tr(W_{ij}^{\star 2}), tr(W_{ij}^{\star 2} b_{jk}^\star), \\ & tr(W_{ij}^{\star 2} b_{jk}^{\star 2}), tr(W_{ij}^{\star 2} b_{jk}^\star W_{kl}^\star b_{lh}^{\star 2}) \;] \,. \end{aligned} \tag{3}$$

Note that in (2) only the mean rotation term W_{ij} has been retained since for the case of frame rotation to be considered below the mean shear S_{ij} is irrelevant.

3 Modelling of the Diffusion-Term

We consider the problem of turbulence generation by a vibrating grid in the $x_2 - x_3$ plane with no mean velocity and rotating about x_1. Analyzing the multi-point correlation equations for this flow with the help of Lie-group methods we find the following set of invariant solutions (see [2]):

$$\begin{aligned} \tilde{x} &= (x_1 + x_o)e^{t/t_o}, \quad \overline{u_i'u_j'}(x_1,t) = \Omega_1^2(x_1+x_o)^2\widetilde{\overline{u_i'u_j'}}(\tilde{x}_1), \\ K &= \Omega_1^2\tfrac{1}{2}(x_1+x_o)^2\widetilde{\overline{u_i'u_i'}}(\tilde{x}_1), \quad \epsilon = \Omega_1^3(x_1+x_o)^2\tilde{\epsilon}(\tilde{x}_1). \end{aligned} \tag{4}$$

The surprising result for the present case is that even for $t \to \infty$ the turbulent diffusion only influences a finite domain due to the quadratic behavior of the Reynolds stress tensor in (4). This case can not be modelled at all by one-point models of classical form. For the correct prediction of this behavior the classical LRR model constants must satisfy the equation

$$114C_\epsilon C_{10}C_sC_{\epsilon 2} - 9C_{10}^2C_\epsilon^2 + 18C_{10}C_\epsilon^2 + 135C_\epsilon C_sC_{\epsilon 2} - 216C_s^2C_{\epsilon 2}^2 = 0, \tag{5}$$

in order to ensure a quadratic Reynolds stress tensor. Still this is rather artificial since frame rotation does enter the model equations. Since (5) is not satisfied by the standard model constants it is necessary to modify them to model the given flow case properly. The modelling of the diffusion-term as well as the modelling of the pressure strain will account for the new solutions.

4 Investigation of the Finite Domain Diffusion Solution

Classical Reynolds-stress transport models augmented by the nonlinear-scrambling-term have been investigated for their capability to account for system rotation for the given flow geometry. In the following the Launder-Reece-Rodi model (LRR) [1] is used as an example.

4.1 Model Equations

Introducing the nonlinear-scrambling-term, the model equations for the Reynolds-stresses and the dissipation respectively are

$$\frac{\partial \overline{u_i'u_j'}}{\partial t} = C_4K\left(b_{ik}W_{jk} + b_{jk}W_{ik}\right) - C_{10}b_{ij}\epsilon - \frac{2}{3}\delta_{ij}\epsilon + \epsilon\phi_{ij}^{nls^\star} \tag{6}$$

$$+ C_s\frac{\partial}{\partial x_k}\left(\frac{K}{\epsilon}\left(\overline{u_i'u_l'}\frac{\partial \overline{u_j'u_k'}}{\partial x_l} + \overline{u_j'u_l'}\frac{\partial \overline{u_k'u_i'}}{\partial x_l} + \overline{u_k'u_l'}\frac{\partial \overline{u_i'u_j'}}{\partial x_l}\right)\right) \tag{7}$$

$$- 2\Omega_k\left[e_{kli}\overline{u_j'u_l'} + e_{klj}\overline{u_i'u_l'}\right], \tag{8}$$

$$\frac{\partial \epsilon}{\partial t} = -C_{\epsilon 2}\frac{\epsilon^2}{K} + C_\epsilon\frac{\partial}{\partial x_k}\left(\frac{K}{\epsilon}\overline{u_k'u_l'}\frac{\partial \epsilon}{\partial x_l}\right) \tag{9}$$

with

$$W_{ij} = \frac{1}{2}\left(\frac{\partial \overline{u_i}}{\partial x_j} - \frac{\partial \overline{u_j}}{\partial x_i}\right) + e_{mji}\Omega_m. \tag{10}$$

These equations are further simplified by the homogeneity in the $x_2 - x_3$ plane and the fact that there are no shear stresses occurring in the given flow. Therefore $\overline{u_2'u_2'}$ becomes equal to $\overline{u_3'u_3'}$ and all off-diagonal elements of the Reynolds-stress tensor become zero.

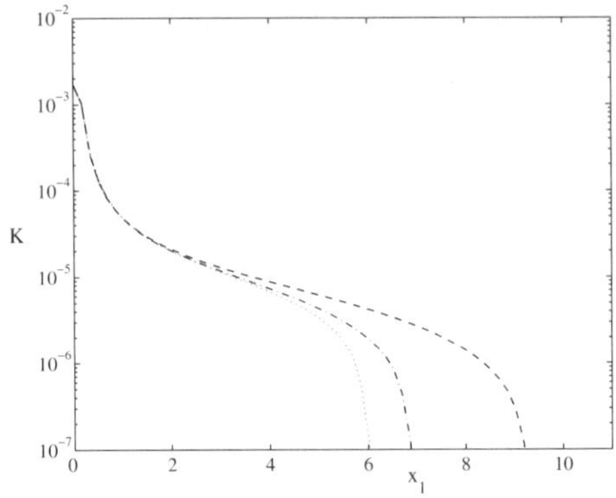

Fig. 2. Dependence of the decreasing behavior on the rotation rate: $---\ \Omega = 0.5$, $-\cdot-\ \Omega = 1$, $\cdots\ \Omega = 2$.

4.2 Results

Taking a closer look at the model equations it turns out that the rapid term of the pressure-strain-correlation as well as the advection term do not contribute to the solutions for the given flow geometry. Therefore only the *nls* term accounts for rotation.

The LRR model augmented by the *nls* term has been investigated numerically for the given flow geometry with the help of a numerical tool, called 1D solver (© S. Wallin, FOI). Thereby the coefficients in the *nls* term have all been put to zero except for β_2 which has been set to 0.5. It thus has been found that as a consequence the *nls* term a finite domain diffusion is predicted. As a result the turbulent kinetic energy K decreases to zero at a finite value of x_1, later called fixed point. However what can not be predicted by the model is the quadratic decreasing behavior and the constant integral timescale which are given by the invariant solutions. Assuming the quadratic decreasing behavior one receives solutions for the Reynolds-stresses which are not realizable with the standard model constants. Therefore these solutions can not be obtained using physical reasonable boundary conditions.

In addition it is found that the position of the fixed point depends on the rotation rate as can be seen from figure 2. The higher the rotation rate the closer the fixed point lies towards the grid.

References

1. Launder B. E., Reece G. C., Rodi W. (1975) J. Fluid Mech. **68**: 537-566
2. Oberlack, M. & Guenther, S. (2003) Fluid Dyn. Res. **33** 453-476.
3. Sjögren T., Johansson A.V. (2000) Phys. Fluids **12**(6): 1554-1572

Predicting Probability for Stochastic Processes with Local Markov Property

Holger Kantz[1], Detlef Holstein[1], Mario Ragwitz[2], and Nikolay K. Vitanov[3]

[1] Max Planck Institute for Physics of the Complex Systems, Noethnitzer Str. 38, 01187 Dresden, Germany kantz@mpipks-dresden.mpg.de

[2] Fraunhofer Institute for Systems and Innovation Research, Breslauer Str. 48, 76139, Karlsruhe, Germany

[3] Institute of Mechanics, Bulgarian Academy of Sciences, Akad. G. Bonchev Str., Block 4, 1113, Sofia, Bulgaria
vitanov@imbm.bas.bg;vitanov@mpipks-dresden.mpg.de

Summary. In this paper we describe a new method for predicting PDFs of observable quantities driven by stochastic processes with a local Markov property. The method deals with large class of nonstationarities by overembeding the vector in the conditional part of the conditional probabilities of the Markov chain which approximates the Markov process. This allows an application of a Farmer-Sidorowich-like prediction scheme [1] in the obtained vector space. Thus the conditional PDF of the investigated quantity for the next time step can be estimated and various forecasts can be performed. As an illustration the method is applied to the problem for the short-term prediction of turbulent wind gusts which are the major danger for the safe operation of wind energy turbines. Predicted gusts can be made innocent by a simple change of the pitch angle of the rotor blades. Within a prediction horizon of few seconds which is sufficient for this purpose the discussed method produces meaningful results.

Stationary time series data with a dominant low-dimensional deterministic component can be analysed and predicted by the methods of the nonlinear time series analysis based on the concept of time delay embedding [2, 3, 4]. Following this concept we can convert a sequence of equidistant in time scalar time series measurements into a sequence of m-dimensional vectors, composed of successive time series elements, $\mathbf{s} = (s_n, s_{n-1}, s_{n-2}, \ldots, s_{n-m+1})$. For purely deterministic low dimensional data, two such vectors at successive times are related to each other by a unique deterministic map, if $m > 2D_f$, where D_f is the dimension of the attractor on which the dynamics happens, and the map can be extracted from the observed dynamics of the neighbouring points in the reconstructed phase space.

Two reasons prevent the wide application of the time-delay embedding concepts to field recorded time series: In many systems outside the laboratory the stationarity is violated because parameters drift or fluctuate during the time of observation, and many observations do not represent a low-dimensional deterministic system. We have to find another way to analyse and predict nonstationary stochastic time series. Let us concentrate our efforts on phenomena that are driven by some unknown vector valued Markov process. Our choice is determined by the fact that in many practical situations the aperiodic time evolution is governed by deterministic nonlinear process coupled with stochastic inputs to the system. An idealised physical description would then consist of a vector valued Langevin equation [5, 6] .

For the investigated time series $\{v_n\}$ we assume that the measurements v_n are obtained by a time-discrete sampling of an instantaneous but otherwise arbitrary projection of the state vectors $\mathbf{x}(t_n)$ of the above Markov process. We do not have any knowledge of this Markov process, and in addition, we cannot expect that the recorded observable represents a one variable continuous state Markov chain of finite order [6] ,even worse, in general we must expect presence of infinite memory. Fortunately, in many situations correlations decay fast with the time lag and for these cases we can approximate the stochastic dynamics of v by a continuous state Markov chain, whose order m is a model parameter we have to choose. We can estimate the probability one step in the future by the following steps

1. The present state of the underlying Markov process $\mathbf{x}(t_n)$ is replaced by the last m measurements, i.e. the probability to observe v' in the next measurement is approximated by $p(v_{n+1} = v' \mid v_n, v_{n-1}, \ldots, v_{n-m+1})$. The vector $(v_n, v_{n-1}, \ldots, v_{n-m+1}) = \mathbf{v}_n$ is formally identical to a delay vector used in the time-delay embedding approach [2, 3]
2. We assume that a slow time dependence of the conditional probability $p(v_{n+1} \mid \mathbf{v}_n)$ can be traced back to the assumption of parameters $\mathbf{a}(t)$ which have to be included as additional conditions. In general $\mathbf{a}(t)$ are typically unknown and we replace them by an appropriate increase of the dimension m of the conditioning vector, which is called overembeding [7], $p(v_{n+1} \mid \mathbf{v}_n, \mathbf{a}(t)) \approx p(v_{n+1} \mid v_n, v_{n-1}, \ldots, v_{n-m-k+1})$. The overembeding holds for stochastic Markov chain modelling only approximatively, i.e., despite the fact that formally infinitely many additional time lags have to be included, we assume that a finite number yields a good approximation [7, 8]. This assumption is well justified for the cases when quantities responsible for non-stationarity, such as surface and air temperatures or pressures, fluctuate at low amplitudes on long time scales.
3. We choose a value m of the number of time steps representing the memory, which is assumed to include already extra k conditions needed to compensate the non-stationarity.
4. We assume that similar conditions give rise to a similar probability distributions and hence the conditional probabilities are smooth in their arguments $\mathbf{v}_n$. The current state vector $\mathbf{v}_n$ is a point in an m dimensional space. Neighbouring points in this space represent similar state vectors and the observed "futures" v_{k+1} of close neighbours $\mathbf{v}_k$ form a random

sample of the distribution $p(v_{n+1} \mid \mathbf{v}_n)$. A histogram of this sample gives a coarse grained view of the estimate of this distribution.

5. The histogram is obtained in the following way. We denote by $\Phi_\epsilon(\mathbf{v}_n)$ a neighbourhood of small diameter ϵ around the vector $\mathbf{v}_n$. The number of vectors $\mathbf{v}_k$ in this neighbourhood taken from the past of the time series, $k < n$, is denoted as $|\Phi_\epsilon(\mathbf{v}_n)|$. For these vectors we inspect the future values v_{k+1} and denote the number of these values which are in the interval $[v', v'+\Delta v']$ as $N(v', \Delta v')$. Then an estimate of the conditional probability at the first future step is

$$p(v' \mid \mathbf{v}_n)\Delta v' \approx \frac{N(v', \Delta v')}{|\Phi_\epsilon(\mathbf{v}_n)|} \tag{1}$$

By means of $p(v' \mid \mathbf{v}_n)$ we can perform various predictions for the investigated time series. Here are two examples

- The optimal prediction $\hat{v}$ of v_{n+1}, in the maximum likelihood sense, i.e., which in an ensemble average minimises the root mean square prediction error, is given by the first moment of this estimated conditional probability,

$$\hat{v}_{n+1} = \int dv'\ v' p(v' \mid \mathbf{v_n}) \approx \frac{1}{|\Phi_\epsilon(\mathbf{v}_n)|} \sum_{k \in \Phi_\epsilon(\mathbf{v}_n)} v_{k+1} \,. \tag{2}$$

- Prediction of the probability of an increment $\Delta v_{n+1} = v_{n+1} - v_n$ larger than g in the next time step is given by the following integral of the probability distribution,

$$p(\Delta v_{n+1} > g) = \int_{v_n+g}^{\infty} dv'\ p(v' \mid \mathbf{v}_n) \tag{3}$$

We illustrate the method by its application for predictions of the probability of a turbulent wind gust to arrive at a measurement device, i.e., the probability that the wind velocity exceeds its current value by more than g m/s during the next 2 seconds. This prediction gives the possibility to make such gusts innocent by a rather small correction of the pitch angle of the rotor blades of the commercial wind turbines, once the occurrence of the gust is predicted. Since wind speeds are correlated, predictions referring to time intervals covering h time steps, $h > 1$, require a suitable generalisation of $p(v'|\mathbf{v_n})$. It is straightforwardly given by the relative number of ϵ-neighbours $\mathbf{v}_k$ whose future fulfils our criterion to be a gust, i.e., whose maximum wind velocity v_{max} between the time steps k and $k+h$ exceeds $v_k + g$ (other gust criteria can be used as well). As a result, for every time step n we create a prediction $\hat{p}_n^{(\text{gust})}$ of the probability of a turbulent gust to occur within the following time window of h time steps, which could be used for the adjustment of the pitch angle of the rotor blades.

As a practical application let us generate a "warning" at every time when the predicted probability exceeds a critical value, $\hat{p}_n^{(\text{gust})} > p_c$, where we choose

$p_c = 0.25$. In Table 1 we report the relative number of warnings thus found for about 50 000 predictions, sorted according to the actual maximum increase of the wind speed. Evidently, with about 15 % false alarms we are able to eliminate about 50 % of the gust events, and the stronger gusts are predicted at a higher rate.

Table 1. The 53085 predictions made on a data set of one day are sorted according to the true $g \leq v_{max} - v_n < g + 1$ during the prediction horizon of $h = 12$ steps (=1.5 s) (first column). The second column reports the number of such events, the third the absolute and the last the relative number of gust-warnings obtained in each subsample. Larger p_c gives a better prediction rate of gusts but introduces also more false alarms.

Gust strength g [m/s]	Number of events	Number of warnings	In percent
0	42181	6342	15%
1	8905	3134	35%
2	1651	807	49%
3	297	170	57%
4	49	30	61%
5	2	1	50%

Finally we note that the discussed method exploits nonlinear higher order temporal correlations in stochastic data without computing them explicitly. Predicting probabilities instead of mean values enables us to make meaningful forecasts of turbulent time series. This scheme can be employed for all data obtained from any stochastic process with local Markov property.

We thank to the Alexander von Humboldt Foundation and to NCSR of Ministry of Education and Science of Bulgaria, contract # MM 1201/02 (N.K.V.) and to the German Ministry of Economy (M. R.) for the support of our research.

References

1. Farmer J. D., Sidorovich. J. J. (1987) Phys. Rev. Lett. 59: 845–848.
2. Takens F. (1981) Lectiture Notes in Mathematics 898: 366–381
3. Sauer T., Yorke J., Casdagli M. (1991) J. Stat. Phys. 65: 579–616
4. Kantz H., Schreiber T. (1997) Nonlinear Time Series Analysis . Cambridge University Press, Cambridge
5. Risken H. (1989) The Fokker-Planck Equation. Springer, Berlin
6. van Kampen N. G. (1992). Stochastic Processes in Physics and Chemistry. North-Holland, Amsterdam
7. Hegger R., Kantz H., Matassini L., Schreiber T. (2000) Phys. Rev. Lett. 84: 4092–4095
8. Kantz H., Ragwitz M. Int. J. Bif. Chaos (in the press).

Stability of Turbulent Kolmogorov Flow

Bernard Legras[1] and Barbara Villone[2]

[1] Laboratoire de Météorologie Dynamique, 24, rue Lhomond F-75231 Paris legras@lmd.ens.fr

[2] IFSI-CNR, corso Fiume, 4 I-10133 Torino villone@to.infn.it

1 Introduction

The instability of the Kolmogorov flow $U = \sin y$ has received a large interest in the literature (see [8, 11, 2, 3, 7, 1, 12] and references therein). This flow exhibits a large-scale instability of the negative viscosity type for Reynolds $Re < \sqrt{2}$. For slightly supercritical conditions, the perturbation follows Cahn-Hilliard equation characterized by an inverse cascade of metastable states with scale growing in time. This cascade involves merging of jets until the gravest mode is reached (or without limit in an unbounded system). It can be halted by adding friction or dispersive effects like Rossby waves to generate a stable solution with multiple alternated jets [5, 7, 6]. Such stabilizing mechanisms have been advocated to explain features observed in the atmosphere of fast rotating Jovian planets and in numerical simulations of turbulence on a rotating sphere [9, 4].

The theory, however, is valid at very low Re while geophysical fluids are characterized by very large Re. Finding rigorously large-scale instabilities at large Re is a formidable task, probably out of reach at the moment. In this paper we address the problem of Kolmogorov flow instability when molecular viscosiy is replaced by one of the popular parameterization for small-scale turbulence representing the motion at scales smaller than the Kolmogorov flow. This is clearly a non rigorous approach, but it provides hints on large-scale instabilities at large Re and, hopefully, on the character of such instabilities.

2 The Stability of a Turbulent Kolmogorov Flow in a Clark-Smagorinsky Model

As in the standard problem [11], we use the framework of the two-dimensional incompressible Navier-Stokes equation. Introducing the streamfunction ψ such that $(u = \partial_y \psi, v = -\partial_x \psi)$, our basic equation is

$$\partial_t \nabla^2 \psi - \frac{\partial(\psi, \nabla^2 \psi)}{\partial(x,y)} = -r \nabla^2 \psi + D + F \tag{1}$$

where r is a friction and the dissipation D follows the Clark-Smagorinsky model [10] often used in LES or atmospheric simulations:

$$\begin{aligned} D = & \frac{\Delta^2}{12}\left((\partial_{xy^3}\psi + \partial_{x^3y}\psi)(\partial_{xx}\psi - \partial_{yy}\psi) + \partial_{xy}\psi(\partial_{y^4}\psi - \partial_{x^4}\psi)\right) \\ & + C_s \Delta^2 \partial_{xy}\left(4(\bar{S}\partial_{xy}\psi) + (\partial_{yy} - \partial_{xx})\bar{S}(\partial_{yy}\psi - \partial_{xx}\psi)\right) \end{aligned}$$

with $\bar{S} = \sqrt{4(\partial_{xy}\psi)^2 + (\partial_{xx}\psi - \partial_{yy}\psi)^2} + \nu$. In this model Δ depends on the filtering of the velocity and can be considered as a cutoff scale. The forcing F is chosen in order to maintain $\Psi(y) = A\cos y$ with $A > 0$ as a stationary solution of (1). In order to distinguish our case from the standard case where this flow is maintained against molecular diffusion, we call it the *turbulent* Kolmogorov flow. A small diffusion ν is added to regularize the solutions near $y = \pm\pi/2$

We assume scale separation between the Kolmogorov flow and the large-scale flow, and introduce slow variables $X = \epsilon x$ and $T = \epsilon^2 t$. Then it turns out that the flow depends only on (X, y, T) and we expand ψ as

$$\psi(X,y,t) = \Psi(y) + \phi_0(X,T) + \epsilon\psi_1(X,y,t) + \epsilon^2\psi_2(X,y,t) + \dots \tag{2}$$

Equation (1) is then expanded in ϵ and the perturbation problem is solved at successive orders. At each order n in the expansion we have to solve

$$C_s \Delta^2 \partial_{yy}((2A|\cos y| + \nu)\partial_{yy}\psi_n - r\psi_n) = \mathcal{H}\,, \tag{3}$$

where $\mathcal{H}$ holds for a complicated expression involving solutions to lower order equations in the perturbation expansion. In order to satisfy (3), the integral $\int_0^{2\pi} \mathcal{H} dy$ must vanish for all (X, T), thus providing solvability conditions for each n.

Since the dissipation is a nonlinear function of the flow, the algebra of the perturbative expansion is considerably more intricate than in the standard problem. In practice it must be solved by symbolic calculations using Mathematica. The output of these calculations would fill several printed pages. The integrals appearing in the solvability conditions do not need to be calculated except for those involved in the amplitude equation below. The numerous other terms are found to vanish by the application of simple symmetry rules.

3 Results and Discussion

It turns out that the solvability conditions at order 0 and 1 are automatically satisfied by (2). At order 2, we get a solvability condition

$$r = \frac{4AC_s\Delta^2}{\pi}\,,$$

that is imposed to get rid of a spurious instability entirely due to the Clark-Smagorinsky parameterization (i.e. without any coupling with the Jacobian in (1)). This effect should be taken into account in numerical simulations of the Kolmogorov flow.

The solvability condition is again satisfied at order 3 and, like in the standard problem, the solvability condition at order 4 provides the instability condition for ϕ_0. This equation is

$$\partial_{X^2T}\phi_0 = H\partial_{X^4}\phi_0 + G(\partial_X\phi_0)^2\partial_{X^2}\phi_0\,, \tag{4}$$

where H and G are two constants depending on the parameters (C_s, Δ, ν) and on the solutions of the perturbation problem at orders 1 and 2.

It turns out that in the limit $\nu \to 0$

$$H = \frac{2\Delta^2 AC_s}{\pi} - \frac{\alpha}{2\Delta^2 C_s}\left(1 + \frac{\Delta^2}{12}\right)A\,,$$

where the coefficient α is numerically calculated as $\alpha = 0.18935\ldots$. If one further take the usual value 0.23 for C_s, it is found that $H < 0$ for $0 < \Delta < \Delta_c = 1.341\ldots$. Hence (4) exhibits negative viscosity for small enough cutoff scale, a result that is very similar to the standard Kolmogorov instability.

The new feature here is that the amplitude equation provides an additional nonlinear term that is absent in the standard problem where Cahn-Hilliard equation appears at sixth order. In the limit $\nu \to 0$, the coefficient in front of this term is

$$G = \frac{1}{AC_s\Delta^2}\left(\alpha_1 + \left(\frac{\alpha_2}{C_s} + \frac{\alpha_3}{C_s^2}\right)\frac{1}{\Delta^2} + \left(\frac{\alpha_4}{C_s^2} + \frac{\alpha_5}{C_s^3}\right)\frac{1}{\Delta^4} + \frac{\alpha_6}{C_s^3\Delta^6}\right)$$

with numerical coefficients $\alpha_1 = 0.012\ldots$, $\alpha_2 = -0.0052\ldots$, $\alpha_3 = -0.0029\ldots$, $\alpha_4 = -0.0026\ldots$, $\alpha_5 = 0.00073\ldots$ and $\alpha_6 = 0.0084\ldots$. It turns out that for $\Delta = \Delta_c$, we have $G \approx 0.219/A$, that is $G > 0$.

In the instability range, (4) admits stationary solutions that can be obtained in terms of elliptical functions. The dynamical study of (4) will be presented elsewhere.

We have shown that a large-scale instability of the Kolmogorov flow is obtained when a parameterized turbulent viscosity replaces the standard molecular viscosity. This result is reached to the price of two scale separation hypothesis, the first one between the Kolmogorov flow and the large-scale flow, and the second one between the small-scale turbulence represented by the Clark-Smagorinsky parameterization and the Kolmogorov flow. Nevertheless, the result is encouraging for the generic existence of large-scale instabilities of the negative viscosity type in fully turbulent flow. The robustness would need to be tested with a variety of parameterizations and numerical simulation. Our results also shows the necessity of damping spurious instabilities that are purely generated by the parameterization and may have polluted previous numerical investigations. We also find a new nonlinear term in the amplitude equation which is obtained without any assumption of slight supercriticality. The term is cubic but differs from the nonlinearity in the Cahn-Hilliard equation. Again, robustness and significance of this result requires further studies. This work has been supported by Project N.11397 CNR/CNRS 2002-2003

References

1. Balmforth N, Young Y (2002) J Fluid Mech 450:131–167
2. Dubrulle B, Frisch U (1991) Phys Rev A 43:5355–5364
3. Frisch U, Legras B, Villone B. (1996) Physica D 94:36–56
4. Huang HP, Robertson WA (1998) J Atmos Sci 55:611–632
5. Legras B, Frisch U, Villone B (1999) Phys Rev Lett 82:4440–4443
6. Legras B, Villone B (2003) Physica D 175:139–168
7. Manfroi A, Young W (1999) J Atmos Sci 56:784–800
8. Meshalkin L, Sinai Y (1961) Appl Math Mech 25:1700–1705
9. Nozawa T, Yoden S (1997) Phys. Fluids 9:2081–2093.
10. Pope SB (2000) Turbulent flows, Cambridge University Press
11. Sivashinsky G, (1985) Physica D 17:243–255
12. Woodruff SL, Shebalin JV, Hussaini MY (1999) NASA/CR-1999-209727 ICASE Report No.99-45:1–16

Conditional Moment Closure Based on Two Conditioning Variables

Jorge R. Lozada Ramirez, W. Kendal Bushe, and Andrea Frisque

Department of Mechanical Engineering, University of British Columbia, Vancouver, BC, V6T 1Z4, Canada, frisque@mech.ubc.ca

Abstract

A conditional moment closure approach for modelling turbulent combustion is proposed, based on two conditioning variables. The two conditioning variables used here are mixture fraction and a second conserved scalar a, which is initialized perpendicular to the mixture fraction, such that the two conditioning variables constitute a plane. With this we hope to capture important physical features of turbulent reacting flows which the single conditioning variable approach cannot, including local ignition, extinction and re-ignition due to small scale strain fluctuations. We propose to model the stress tensor with a stochastic process to account for chaotic turbulent fluctuations.

1 Introduction

In models for turbulent combustion, in addition to the challenges in hydrodynamic turbulence, closure for the nonlinear chemical source-term in the species balance equations is needed. The chemical source-term is governed by Arrhenius-type reaction constants with a pre-exponential factor: $k = A \cdot T^B e^{RT/E}$ (A,B,R and E are constants). Therefore the average of the chemical source-term $\overline{\dot{\omega}_r(T, c_j)}$ cannot be approximated by the chemical source-term of the average temperature $\overline{T}$ and the average species concentrations $\overline{c_j}$, $\overline{\dot{\omega}_I(T, c_j)} \neq \dot{\omega}_I(\overline{T}, \overline{c_j})$.

In Conditional Moment Closure (CMC)[5], closure is sought for terms in the transport equations for the conditional moment of scalars (like temperature $\langle T|Z\rangle$). In non-premixed systems the most suitable conditioning variable is the mixture fraction Z, which characterizes the mixedness of the reactants. CMC gives good closure for the chemical source-term, but, when only one conditioning variable and only one conditional moment is used, the model cannot capture important physical phenomena such as ignition, extinction or re-ignition. This is because the temperature can vary significantly along an isopleth of mixture fraction and one cannot capture these fluctuations using only one conditioning variable.

In Fig. 1 a scatter plot of the temperature over mixture fraction from a direct numerical simulation (DNS) of decaying isotropic turbulence with a reduced reaction mechanism representing methane/air combustion [2] is shown, as well as the corresponding conditional moment. There still are significant fluctuations of the actual temperature around the conditional average, particularly towards low temperatures, due to local extinction and re-ignition. These fluctuations can be attributed to the effects of local fluctuations in the rate of strain. In the hopes of capturing such effects, Bushe [1] proposed to use a second conditioning variable to account for variations in local strain along the isosurface of mixture fraction. Introducing a scalar a such that ∇a is orthogonal to ∇Z the conditional average of a scalar Y_I (representing mass fractions and temperature) with two conditioning variables becomes:

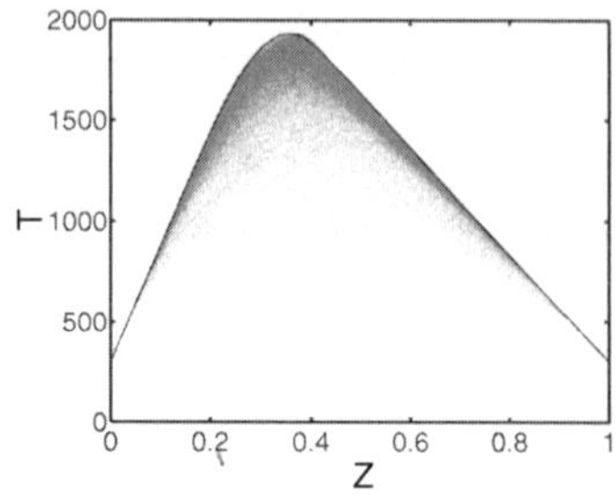

Fig. 1. Scatter plot of the temperature for combustion in decaying turbulence [2].

$$\overline{Y_I} = \overline{Y_I|\zeta,\alpha} = \overline{Y_I}(\zeta,\alpha;x_k,t) \equiv \langle Y_I(x_k,t)|Z(x_k,t)=\zeta, a(x_k,t)=\alpha\rangle$$

2 Model

The basic assumptions in the model are:

- The mixture fraction Z and the scalar a are conserved:

$$\frac{\partial Z}{\partial t} + u_k\frac{\partial Z}{\partial x_k} - D_I\frac{\partial^2 Z}{\partial x_k \partial x_k} = 0\,, \quad \frac{\partial a}{\partial t} + u_k\frac{\partial a}{\partial x_k} - D_I\frac{\partial^2 a}{\partial x_k \partial x_k} = 0$$

- Initially ∇a is orthogonal to ∇Z and it remains so on average.
- $\rho D \approx$ constant.
- For an isotropic, homogeneous flow, all spacial derivatives of conditional averages and their fluctuations vanish.
- $\overline{\left(\frac{\chi_Z}{\chi_a}\right)} \approx \frac{\overline{\chi}_Z}{\overline{\chi}_a}$ and $\overline{\left(\frac{\chi_a}{\chi_Z}\right)} \approx \frac{\overline{\chi}_a}{\overline{\chi}_Z}$, where $\chi_Z = \frac{\partial Z}{\partial x_k}\frac{\partial Z}{\partial x_k}$ and $\chi_a = \frac{\partial a}{\partial x_k}\frac{\partial a}{\partial x_k}$ are the scalar dissipations of mixture fraction and the scalar a, respectively.

Ruetsch and Maxey [6] provided a transport equation for scalar dissipation

$$\frac{\partial \chi_Z}{\partial t} + u_k\frac{\partial \chi_Z}{\partial x_j} = -2\frac{\partial Z}{\partial x_i}\frac{\partial Z}{\partial x_j}S_{ij} - 2D\frac{\partial^2 Z}{\partial x_j \partial x_i}\frac{\partial^2 Z}{\partial x_j \partial x_i} + D\frac{\partial^2 \chi_Z}{\partial x_j \partial x_j}\,, \tag{1}$$

where $S_{ij} = \frac{1}{2}\left(\frac{\partial u}{\partial y} + \frac{\partial v}{\partial x}\right)$ is the strain tensor. Conditionally averaging the transport equation for the scalars Y_i and Eq. 1 under the basic assumptions mentioned before and neglecting the term $\overline{u'_k\frac{\partial \chi'_Z}{\partial x_k}}$ (these are fluctuations around the conditional average) yields the model equations:

$$\begin{aligned}
\overline{\rho}\frac{\partial \overline{Y_I}}{\partial t} &= \overline{\dot\omega}_I + \frac{\partial^2 \overline{Y_I}}{\partial \zeta^2}\rho D\overline{\chi_Z} + \frac{\partial^2 \overline{Y_I}}{\partial \alpha^2}\rho D\overline{\chi_a} \\
\frac{\partial \overline{\chi_a}}{\partial t} &= -2\overline{\frac{\partial a}{\partial x_i}\frac{\partial a}{\partial x_j}S_{ij}} - \frac{D}{2}\left[\left(\frac{\partial \overline{\chi_a}}{\partial \zeta}\right)^2 + \left(\frac{\partial \overline{\chi_a}}{\partial \alpha}\right)^2\frac{\overline{\chi_Z}}{\overline{\chi_a}} - 2\frac{\partial^2 \overline{\chi_a}}{\partial \zeta^2}\overline{\chi_Z} - 2\frac{\partial^2 \overline{\chi_a}}{\partial \alpha^2}\overline{\chi_a}\right] \\
\frac{\partial \overline{\chi_Z}}{\partial t} &= -2\overline{\frac{\partial Z}{\partial x_i}\frac{\partial Z}{\partial x_j}S_{ij}} - \frac{D}{2}\left[\left(\frac{\partial \overline{\chi_Z}}{\partial \zeta}\right)^2 + \left(\frac{\partial \overline{\chi_Z}}{\partial \alpha}\right)^2\frac{\overline{\chi_a}}{\overline{\chi_Z}} - 2\frac{\partial^2 \overline{\chi_Z}}{\partial \zeta^2}\overline{\chi_Z} - 2\frac{\partial^2 \overline{\chi_Z}}{\partial \alpha^2}\overline{\chi_a}\right].
\end{aligned}$$

Only the terms involving the stress tensor are unclosed in these equations.

A Coupled Map Lattice (CML) for the velocity increments in fully developed, isotropic turbulence modelling the energy cascade has been proposed by Hilgers and Beck [4]. Using the following map, realizations of velocity increments from the integral to the dissipation range can be generated.

$$\begin{aligned}
x_{n+1}(i) &= T(x_n(i)), \qquad T(x) = 1 - 2x^2 \\
u^{(1)}_{n+1}(i) &= \lambda_1 u^{(1)}_n(i) + \frac{g}{2d}\left(\sum_\sigma u^{(1)}_n(i+\sigma)\right) + x_{n+1}(i) \\
u^{(k)}_{n+1}(i) &= \lambda_k u^{(k)}_n(i) + \frac{g}{2d}\left(\sum_\sigma u^{(k)}_n(i+\sigma)\right) + C\xi^{(k-1)}_n(i)(1-\lambda_{k-1})u^{(k-1)}_n(i),
\end{aligned}$$

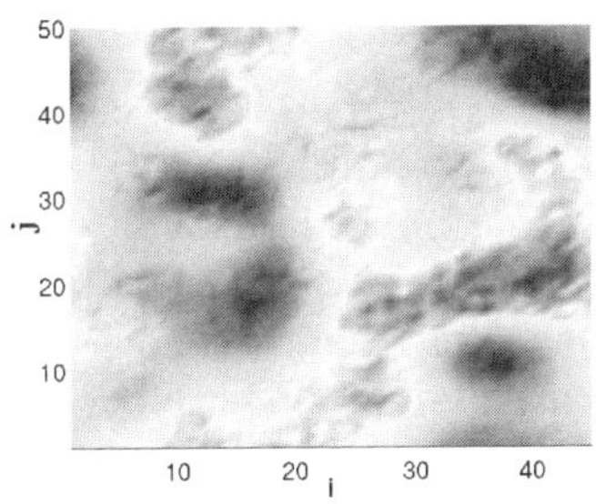

Fig. 2. A realization of the strain S_{ij} with the CML model.

where $k = 2, ..., K$, g is a coupling constant, d the dimension, σ represents the next neighbours, $\lambda_k = e^{-\gamma\tau k}$ a damping constant and $\gamma\tau$ is proportional to the Reynolds number. Two independent processes can be combined to simulate velocity increments in two dimensions, u and v (and three dimensions analogously), from which the stress tensor can be obtained. An example realization of the strain is shown in Fig. 2. Figs. 3(a) and (b) show the resulting fluctuations in χ_Z and χ_a respectively.

We also propose an alternative model for the strain using a periodic force with a random phase shift, based on a heuristic description of the large scales of turbulence similar to the system used to approximate stirring in pseudo-spectral DNS of homogeneous turbulence [3]. The strain resulting from a periodic force with a random phase shift is defined as

$$S_{ij} = \sum_{i=1}^{n}\Psi_n\,, \text{with } \Psi_n = A_n\sin\left(\omega_{n\alpha}\alpha + \theta_{n\alpha}\right)\sin\left(\omega_{n\zeta}\zeta + \theta_{n\zeta}\right)\,,$$

where θ is a random force represented by Gaussian random numbers. This leads to similar strain fields and scalar dissipations as in Figs. 2 and 3 but requires considerably less computational effort.

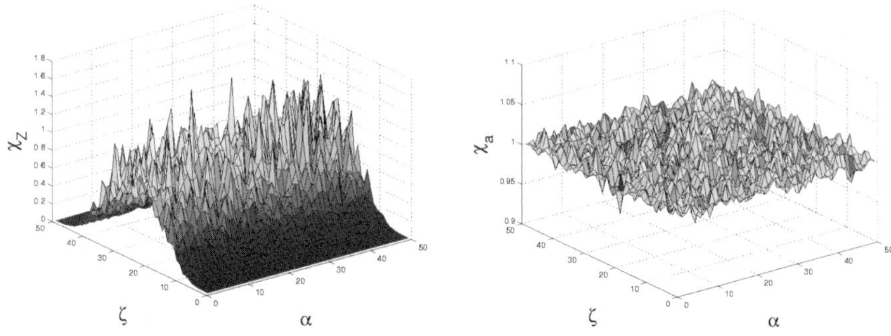

(a) χ_Z (b) χ_a

Fig. 3. Fluctuations in the scalar dissipations.

3 Summary and Conclusions

We have proposed an extension of Conditional Moment Closure including a second conditioning variable to capture the effects of small scale fluctuations in turbulent strain on ignition, extinction and re-ignition phenomena. The small scale strain is modelled successfully using either a Coupled Map Lattice for the simulation of velocity increments in isotropic, homogeneous turbulence or a periodic force model. Further work is ongoing to study the respective properties and to validate the model against DNS results.

References

1. W. K. Bushe. *Conditional Moment Closure Methods for Autoignition Problems.* PhD thesis, University of Cambridge, 1996.
2. W. K. Bushe, R. W. Bilger, and G. R. Ruetsch. Incorporating realistic chemistry into direct numerical simulations of turbulent non-premixed combustion. In *Annual Research Briefs. Center for Turbulence Research, NASA Ames/Stanford University*, page 195, 1997.
3. V. Eswaran and S. B. Pope. Examination of forcing in direct numerical simulations of turbulence. *Comput. Fluids*, 16:257, 1988.
4. A. Hilgers and C. Beck. Hierarchical coupled map lattices as cascade models for hydrodynamical turbulence. *Europhysics Letters*, 45(5):552, 1999.
5. A. Yu. Klimenko and R. W. Bilger. Conditional moment closure for turbulent combustion. *Prog. Energy Combust. Sci*, 25:595, 1999.
6. G. R. Ruetsch and M. R. Maxey. Small-scale features of the vorticity and passive scalar fields in homogeneous isotropic turbulence. *Phys. Fluids A*, 3:1587, 1991.

Stochastic Partial Differential Equations as a Tool for Solving PDF Equations

Vladimir Sabel'nikov and Olivier Soulard

ONERA Chemin de la Hunière 91761 Palaiseau Cedex France

Probability density functions (PDF) present seducing features for modelling turbulent reactive flows: they carry a detailed one-point statistical information and allow to treat chemical source terms exactly. But to be able to take full advantage of those modelling abilities, it is first necessary to possess an efficient numerical method to solve PDF equations.

In this article, this issue is adressed in the frame of Eulerian Monte Carlo methods: stochastic partial differential equations which are stochastically equivalent to the PDF equations are proposed and applied to the calculation of the PDF of a reactive scalar.

1 On Lagrangian and Eulerian Monte Carlo Methods in Turbulent Combustion

One particularity of PDF equations is their potential high dimensionality. As a result, classical methods, such as finite differences, cannot be used, as their computational cost increases exponentially with the number of dimensions.

Up to now, under the impulsion given by the seminal work of Pope [1], PDF equations have mostly been solved using Lagrangian Monte Carlo (LMC) methods: stochastic particles are evolved given prescribed stochastic ordinary differential equations (SODE). The LMC approach yields a computational cost increasing linearly with the number of dimensions and its efficiency has been proven on many different configuration.

However, a few deficiencies have also been observed. In particular, the sampling error is not always controlled with precision, as it depends on the particle distribution in the physical domain. Furthermore, to improve convergence rates, it is often necessary to couple the LMC method to a eulerian RANS solver; this can quickly result in a heavy tool to manipulate, due to the different nature (lagrangian/eulerian) of the solvers [7].

All of the mentionned shortcomings arise from the lagrangian nature of the LMC method. By essence, a Eulerian Monte Carlo (EMC) approach would be devoid of them, and, as any Monte Carlo method, it would keep a computational cost increasing linearly with the number of dimensions. EMC methods are based on stochastic eulerian fields that evolve from prescribed stochastic partial differential equations (SPDE). EMC methods have been extensively used in several domains [8, 9]. However, their application to the simulation of turbulent reactive flows only seems to be dating back from the recent work by Valiño [4].

Despite yielding correct equations, the derivation of the method proposed in [4] seems to be suffering from several inaccuracies which might prejudice its understanding and might limit its range of applicability. Thus, the derivation is based on several restrictive hypothesis: the stochastic fields are required to be smooth and twice differentiable in space. This is not necessary and not consistant with the fact that discontinuities can be created by the equations that are eventually obtained. An attempt is also undertaken to prove that the stochastic fields and the PDF have "similar grade of spatial smoothness". A mapping is introduced to establish this property but the overall proof only holds if the mapped fields are themselves smooth, which remains an hypothesis. Besides, the procedure for obtaining the SPDE governing the stochastic fields does not allow to gain any insight into the obtained equation: this equation is interpreted in [4] as a convection/diffusion equation whereas it can be shown to be an advection equation. Finally, despite being a crucial step, no analysis is carried out to show how to derive a numerical scheme from the SPDE.

In this article, the application of EMC methods to the simulation of turbulent reactive flows is considered under a new angle. This new vision stems from a further interpretation given to the SODES of the LMC methods. As a result, a new path is proposed to derive the SPDEs governing the stochastic fields of EMC methods and a connection is established between Lagrangian and Eulerian methods.

2 A Bridge Between Lagrangian and Eulerian Monte Carlo Methods

Let us consider the Favre averaging PDF f_c governing the one-point statistics of a turbulent reactive scalar c [1]:

$$\begin{aligned} &\frac{\partial}{\partial t}\left(\langle\rho\rangle f_c\right)+\frac{\partial}{\partial x_j}\left(\langle\rho\rangle \tilde{U_j} f_c\right)\\ &=\frac{\partial}{\partial x_j}\left(\langle\rho\rangle \Gamma_T \frac{\partial f_c}{\partial x_j}\right)+\frac{\partial}{\partial c}\left(\langle\rho\rangle\langle\omega_c\rangle(c-\tilde{c}) f_c\right)-\frac{\partial}{\partial c}\left(\langle\rho\rangle S(c) f_c\right) \end{aligned} \tag{1}$$

In this equation, a gradient diffusion hypothesis with coefficient Γ_T is used to model turbulent advection. Micromixing is modelled using the IEM model with scalar frequency $\langle\omega_c\rangle$. $S(c)$ is the chemical source term, $\langle\rho\rangle$ the mean density and $\tilde{U_j}$ is the Favre averaged velocity.

The Fokker-Planck equation (1) is stochastically equivalent to the SODE system with Ito interpretation [1]:

$$\begin{cases} \langle\rho\rangle\, dc = -\langle\rho\rangle \langle\omega_c\rangle (c-\tilde{c})dt + \langle\rho\rangle S(c)dt \\ \langle\rho\rangle\, dx_j = \langle\rho\rangle \tilde{U}_j dt + \frac{\partial}{\partial x_j}\left(\langle\rho\rangle \Gamma_T\right)dt + \langle\rho\rangle \sqrt{2\Gamma_T} dW^j(t) \end{cases} \tag{2}$$

where $W^j(t)$ are independant brownian processes. In LMC methods, the SODEs (2) are directly used to compute of lagrangian particle trajectories.

The key idea that is proposed here is to interpret the SODEs (2) as the stochastic characteristics of the following hyperbolic generalised equation with Stratonovitch interpretation (hereafter indicated with the symbol $\circ$):

$$\begin{aligned} &\langle\rho\rangle \tfrac{\partial c}{\partial t} dt + \left(\langle\rho\rangle \tilde{U}_j - \tfrac{1}{2}\tfrac{\partial}{\partial x_j}\left(\langle\rho\rangle \Gamma_T\right)\right)\tfrac{\partial c}{\partial x_j} dt + \langle\rho\rangle \sqrt{2\Gamma_T}\tfrac{\partial c}{\partial x_j} \circ dW^j(t) \\ &= -\langle\rho\rangle \langle\omega_c\rangle (c-\tilde{c})dt + \langle\rho\rangle S(c)dt \end{aligned} \tag{3}$$

As opposed to LMC methods, this hyperbolic equation is used to compute the evolution of eulerian fields. A proof that equation (3) yields a correct evolution for the PDF (1) has been given in [7], based on [3]. In these derivations, no hypothesis is made about the smoothness, differentiability and length scale of the stochastic field.

Equation (3) can be reformulated with an Ito interpretation:

$$\begin{aligned} &\langle\rho\rangle \tfrac{\partial c}{\partial t} dt + \langle\rho\rangle \tilde{U}_j \tfrac{\partial c}{\partial x_j} dt + \langle\rho\rangle \sqrt{2\Gamma_T}\tfrac{\partial c}{\partial x_j} dW^j(t) - \tfrac{\partial}{\partial x_j}\left(\langle\rho\rangle \Gamma_T \tfrac{\partial c}{\partial x_j}\right) \\ &= -\langle\rho\rangle \langle\omega_c\rangle (c-\tilde{c})dt + \langle\rho\rangle S(c)dt \end{aligned} \tag{4}$$

Despite the presence of a second order spatial operator, equation (4) remains an advection equation: the last two terms on the left hand side are both contributing to the advection of the stochastic field.

3 Numerical Aspects

The numerical analysis focuses on the equation with Ito interpretation (4). Its numerical integration is considered in terms of weak convergence and accuracy.

As already mentionned, this equation is an advection equation and is dealt with as so in the derivation of a numerical scheme. In particular, the scheme is required to be monotone. Besides, because of the Ito interpretation, the numerical scheme should preserve a zero correlation between the Wiener noise and the first order derivative which multiplies it. From these considerations, a second order monotone centered UNO scheme [5] is proposed for the derivative that multiplies the Wiener noise and a simple second order centered scheme is proposed for the diffusion-like term. This eventually yields a monotone advection Lax-Wendroff like scheme for the stochastic part of equation (4). Finally, the deterministic part of equation (4) is treated with traditional techniques.

Temporal integration is adressed by recasting equation (4) in an SODE form [7] which allows the use of traditional SODE techniques [2]. A first order explicit euler scheme is chosen. Thus, the resulting scheme is monotone, explicit, first order in time and second order in space.

The proposed numerical method was tested on simplified monodimensional cases with constant velocity and constant turbulent diffusion coefficient [7]. The orders of spatial and temporal accuracy were checked, as well as the statistical convergence, for both stationary and unstationary problems. The method was then applied to the calculation of a backward facing step in combustion and the results were compared against experimental data. A good agreement was obtained on temperature means and variances [7].

4 Conclusions

A new path is proposed to derive the SPDEs governing the fields of a Eulerian Monte Carlo method. The notion of stochastic characteristic is introduced to obtain hyperbolic SPDEs from the SODEs used in lagrangian approaches. The numerical integration of the corresponding SPDEs is discussed and a numerical scheme is adapted.

The overall procedure is then applied to the calculation of a turbulent reactive scalar PDF. Numerical tests are carried out to check the accuracy of the method. Finally, the simulation of a backward facing step in combustion is performed.

References

1. Pope, S. B., PDF methods for turbulent reactive flows, Progress in Energy and Combustion Science, (1985), 27: 119-192
2. Gardiner, C. W., Handbook of Stochastic Methods,2nd edition, Springer, 1985
3. Eyink, G. and J. Xin, Statistical Analysis if a semilinear hyperbolic system advected by a white in time random velocity field, arXiv:nlin.SI/0201024v1, 2002 (January)
4. Valiño, L., A field Monte Carlo formulation for calculating the probability density function of a single scalar in a turbulent flow, Flow, turbulence and combustion, 1998, 60: 157-172,
5. Levy, D. and G. Puppo and G. Russo, Central WENO schemes for hyperbolic systems of conservation laws, M2AN Math. Model. Numer. Anal., 1999, 33: 547-571
6. Shu, C.-W. and Osher S., Efficient implementation of Essentially Non-Oscillatory shock-capturing schemes II, J. Comp. Phys., 1989, 83: 32-78
7. Soulard, 0. and V. Sabel'nikov, Méthode stochastique eulérienne pour la résolution des équations PDF et application à la simulation des écoulements turbulents réactifs., ONERA/DEFA, 2003 (septembre)
8. Osnes, H. and H. P. Langtangen, A study of some finite difference schemes for a unidirectional stochastic transport equation, SIAM J. Sci. Comput., 1998, 19, (3): 799-812
9. Werner, M. J. and P. D. Drummond, Robust algorithms for solving stochastic partial differential equations, J. Comp. Phys., 1997, 132: 312-326

Non-unique Self-similar Turbulent Boundary Layers in the Limit of Large Reynolds Number

B Scheichl and A Kluwick

Institute of Fluid Mechanics and Heat Transfer, Vienna University of Technology, Resselgasse 3/E322, A-1040 Vienna, Austria bernhard.scheichl@tuwien.ac.at

A rigorous asymptotic analysis concerning the phenomenon of non-uniqueness of quasi-equilibrium turbulent boundary layers in the large Reynolds number limit has recently been carried out in [2]. The approach contains the classical asymptotic theory of wall-bounded turbulent shear flows, cf. [3], as a limiting case. Compared to the latter, the novel theory allows for a moderately large but still asymptotically small velocity defect with respect to the external inviscid flow. Therefore, it applies to attached flow only which, however, exhibits some properties known from separating turbulent boundary layers. Here a first comparison of the theoretical results with numerical and experimental data is presented. As a special aspect, the impact of the equilibrium conditions on the associated external potential flow field is elucidated.

1 Fundamentals and Basic Concepts

Near-equilibrium nominally two-dimensional incompressible turbulent boundary layers play an important role in many internal flow situations. An example is provided by diffuser flows where the boundary layer has to sustain a (preferably) large pressure rise, exerted by the outer irrotational bulk flow. In order to prevent separation it is then advisable to control the pressure gradient such that the boundary layer globally is in an almost self-preserving state, or, equivalently, remains close to equilibrium.

1.1 Classical Small-defect Theory

Let Re denote a suitably defined global Reynolds number. Classical analysis on the basis of the Reynolds equations in the limit $Re \to \infty$, cf. [3], then shows that the main boundary layer characteristics may be expressed as

$$\begin{gathered} 1 - u/U_e = \varkappa(s) + \epsilon F_1'(\eta) + O(\epsilon^2), \quad \eta = y/\delta(s,\epsilon), \\ \delta \propto \epsilon s + O(\epsilon^2), \quad U_e(s) \propto s^m, \quad s = x - x_v. \end{gathered} \tag{1}$$

Here u denotes the streamwise velocity component and U_e its value imposed at the boundary layer edge by the external irrotational free-stream flow. The principal perturbation parameter $\epsilon = O\big((\ln Re)^{-1}\big)$ characterises the magnitude of the boundary layer thickness δ as well as the velocity defect $1 - u/U_e$ in the fully turbulent outer main layer. Furthermore, F_1 is the stream function, and x, y are curvilinear coordinates along and normal to the surface considered, respectively, and x_v denotes the virtual origin. The local curvature of the surface contour is denoted by $\varkappa$. All quantities in (1) are non-dimensional with respect to a suitable global length and velocity scale, respectively.

Furthermore, we note that classical theory establishes the well-known condition necessary for self-similarity in leading order that the inviscid surface slip velocity U_e must vary as a power m of the streamwise coordinate x. Most important, it is argued in [2] by employing first principles that $m > -1/3$ whereas the classical boundary layer structure applies only when $\mu \equiv m + 1/3 = O(1)$.

1.2 New Theory: Distinguished Limit $\mu^{3/2} \ln Re = O(1)$

As pointed out in [2], the breakdown of classical theory associated with the limit $\mu \to 0_+$ is prevented by considering a wake-type flow indicating a state of incipient separation. Then (1) still holds but ϵ is seen to be of $O\big((\ln Re)^{-2/3}\big)$. However, the wall shear remains of $O\big((\ln Re)^{-2}\big)$ as in the classical case, thus entering the analysis in second order only. Assuming quasi-equilibrium implies

$$m + 1/3 \propto \epsilon\hat{\mu}(s) + O(\epsilon^2) \quad \text{and} \quad \varkappa = k(s) + O(\epsilon), \;\; k(s) = \Lambda/s. \tag{2}$$

Herein Λ denotes a free constant with the values $\Lambda = 0$, $\Lambda < 0$, and $\Lambda > 0$ for a plain, concave, and convex surface, respectively. The condition (2) states the remarkable result that the exponent m slightly depends on Re.

Restricting the investigation to flows having $\mathrm{d}\hat{\mu}/\mathrm{d}s \equiv 0$ agrees with the requirement that the boundary layer is in equilibrium up to second order. The necessary condition derived from the second-order integral momentum balance without employing any turbulence closure represents one of the main results of the analysis. Written in the canonical form (dashed curve in Fig. 1 a)

$$9\hat{D}^2\hat{\mu} = 1 + \hat{D}^3 \tag{3}$$

it provides a relationship between $\hat{\mu}$, the rescaled measure of the velocity defect $\hat{D} = O(1)$, and the wall shear stress which is scaled to 1. In the case $\Lambda = 0$ the quantity $\hat{D}^3$ is directly proportional to the shape factor G, see [3], if the contributions of Reynolds normal stresses are neglected. Most interestingly, (3) gives a theoretical explanation for the early experimental observations of non-unique near-equilibrium flows for a given value of m, [1]. The solid curves in Fig. 1 (a) refer to numerical solutions of the boundary layer equations for finite values of Re having adopted a simple mixing-length shear stress closure, see [2]. Due to the logarithmic dependence of ϵ on Re convergence for $Re \to \infty$ to the limit given by (3) is rather slow. In Fig. 1 (b) these results are compared with measurements of a flow approaching separation where m varies slowly.

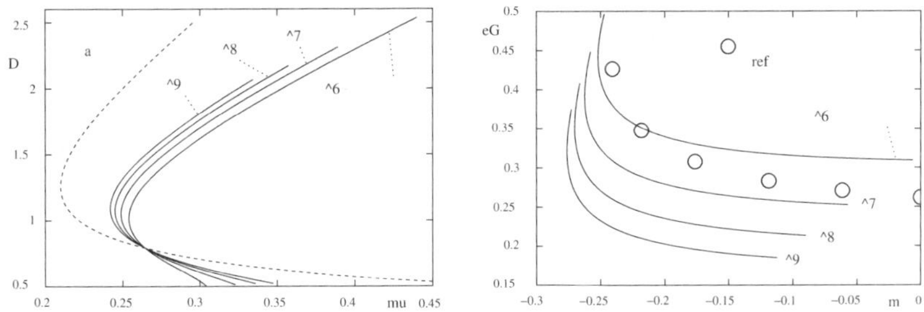

Fig. 1. (*a*) Canonical defect measure $\hat{D}$, (*b*) data for G by [4] taken from [3]

2 Effect of Surface Curvature on External Bulk Flow

Outside the boundary layer the flow is inviscid and irrotational to the order considered here. The latter property is expressed by the Laplace equation, satisfied by the stream function $\psi(x, y)$. If, as before, the flow is taken to be in equilibrium, the slip velocity U_e at the solid surface is given by (1). Hence,

$$\begin{gathered} \partial_s(h^{-1}\partial_s\psi) + \partial_y(h\,\partial_y\psi)_y = 0, \\ h = 1 + \varkappa(s)\,y, \quad y = 0\colon\ \ \psi = 0,\ \ \partial_y\psi = U_e(s) \propto s^m. \end{gathered} \tag{4}$$

Supplementing (4) with appropriate inflow and outflow as well as boundary (symmetry) conditions prescribed at the opposite wall (the centerline) of a duct, for example, then defines a well-posed elliptic problem determining ψ. In general, one expects also the wall curvature $\varkappa(s)$ to be part of the solution.

In contrast, equilibrium in the limit $m + 1/3 \to 0_+$ considered here imposes the additional condition for $\varkappa(s, Re)$ provided by (2). Note that by definition $\varkappa \equiv \mathrm{d}\phi/\mathrm{d}s$, see Fig. 2 (*b*). Let X, Y denote Cartesian global coordinates such that, without any loss of generality of the analysis, $s = 1$ and $\phi = 0$ at the origin $X = 0$, $Y = 0$. By integration of the second relation in (2) one then obtains a representation $X = X_c(s; \Lambda)$, $Y = Y_c(s; \Lambda)$ of the surface,

$$\begin{gathered} X_c = \big(s(\Lambda\sin\phi + \cos\phi) - 1\big)/\lambda, \quad Y_c = \big(s(\Lambda\cos\phi - \sin\phi) - \Lambda\big)/\lambda, \\ \phi = \Lambda\ln s, \quad \lambda = 1 + \Lambda^2. \end{gathered} \tag{5}$$

For $\Lambda \neq 0$ the spiral-type curves defined by (5) have their center in the singular point S of the flow field where $s = x - x_v = 0$ and $y = 0$, see Fig. 2 (*b*). In this case, therefore, the equilibrium flows associated with these contours given by (5) are assumed to take place sufficiently far downstream of $s = 0$.

Combining the specific similarity structure of the boundary layer, see (1), and the condition in (2) for the wall curvature for any value of m suggests the existence of self-similar solutions of (4) of the form $\psi \propto s^{m+1}g(\zeta; \Lambda)$, $\zeta = y/s$. Indeed, inserting this ansatz into (4) yields an ordinary initial value problem for g. In the inviscid limit $m = -1/3$ the problem (4) is then rewritten as

$$\begin{gathered} h(h^2 + \zeta^2)g'' + \big(\Lambda(h^2 - \zeta^2) + 2\zeta h/3\big)g' + (4\Lambda\zeta/9 - 2/9)g = 0, \\ h = 1 + \Lambda\zeta, \quad \zeta = 0\colon\ \ g = 0,\ \ g' = 1. \end{gathered} \tag{6}$$

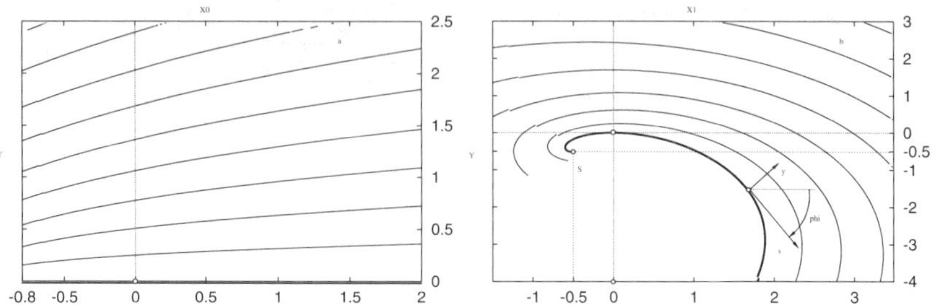

Fig. 2. Streamlines $\psi = 0, 0.25, 0.5, \ldots$ and wall contour (bold): (*a*) $\Lambda = 0$, (*b*) $\Lambda = 1$

Here primes denote derivatives with respect to ζ. The problem (6) admits the closed-form solution $g = 3/2\,(1+\zeta^2)^{1/3}\sin(2/3\,\arctan\zeta)$ which describes the potential flow past a flat wall, i.e. $\Lambda = 0$, but has to be solved numerically in general. Two representative results are depicted in Fig. 2 after applying the transformation $X = X_c + y\sin\phi$, $Y = Y_c + y\cos\phi$.

3 Conclusions and Further Outlook

An asymptotic analysis of turbulent equilibrium flows in the double limit $Re \to \infty$, $m + 1/3 \to 0_+$ has been presented. The results obtained appear to be supported by existing experimental evidence. It has to be noted, however, that experimental data which satisfy the theoretical requirements in a strict sense appear not to be available at present, so that a more rigorous test of the theoretical predictions has to wait for further experimental and/or numerical efforts. In this connection we add that the potential flow solution for a flat wall, i.e. $\Lambda = 0$, fixes a suction rate given by $-\partial_x\psi$ at an opposite wall $Y = const$ to be used in an experimental setup or in order to perform LES/DNS.

We also note that the solutions of (6) in the case $\Lambda > 0$ may be of engineering relevance. For example, consider a spiral diffuser duct formed by contours which collapse onto the convex curve $y = 0$ and a further potential flow streamline where $y > 0$, respectively. The boundary layer adjacent to its convex inner wall withstands the maximum pressure rise possible for a given Reynolds number. Although the opposite boundary layer on the concave outer wall is clearly not in equilibrium it is exposed to a weaker pressure gradient (although over a longer distance) and thus has a smaller velocity defect. Hence, it is supposed to be less sensitive to separation, but this remains to be shown.

This work was supported by the Austrian Science Fund (FWF) under grant number P 16555-N.12.

References

1. Clauser FH (1954) J Aeronaut Sci 21:91–108
2. Scheichl B, Kluwick A (2004) J Fluid Mech, submitted
3. Schlichting H, Gersten K (2000) Boundary-layer theory. Springer, Berlin
4. Simpson RL, Chew Y-T, Shivaprasad BG (1981) J Fluid Mech 113:23–51

Part IV

Experiments

Two-Point-Correlations in a Zero Pressure Gradient Boundary Layer at $Re_\theta = 54600$

R. Abstiens, W.A. El-Askary, and W. Schröder

Aerodynamisches Institut RWTH Aachen, Wüllnerstr. zw. 5 u. 7, 52062 Aachen, Germany, ronald@aia.rwth-aachen.de

Summary. The measurements were performed in the $8 \times 6\ m^2$ test section of the low-speed German-Dutch wind tunnel (DNW-LLF) at Reynolds numbers up to $Re = 50 \times 10^6$ in a zero pressure gradient turbulent boundary layer with a maximum thickness of 150 mm. Two triple hot-wire probes were used, one of which could be shifted in the vertical direction and the other one could be moved such that geometrically a three-dimensional wedge-like zone above the surface of the flat plate could be analyzed. Furthermore, 2C and 3C Particle-Image Velocimetry (PIV) measurements were done to obtain the instantaneous flow structure in a complete plane and to check the validity of fundamental assumptions such as the Taylor hypothesis.

1 Introduction

Publications over the last fifteen years, e.g. [1] have stimulated the discussion concerning the validity of the log-law and its constants of the normalized mean velocity distribution especially in the high Reynolds number regime. Since such flows are still out-of-reach for direct numerical simulations, reliable experiments are required to obtain a deeper understanding of the details of a turbulent boundary layer at high Reynolds numbers. In 1994 [2] the mean velocity and Reynolds stress profiles were measured at high Reynolds numbers in the incompressible turbulent boundary layer with zero pressure gradient on the smooth side wall of the DNW-LLF in the Netherlands. These measurements provided a complete set of data of the turbulent stresses for Reynolds numbers based on the momentum thickness up to $Re_\theta = 60000$. To ensure the accuracy of the measurements two different hot-wire probes were simultaneously used and placed next to each other in the experimental set up. The reason for the investigations discussed in this study is the undefined boundary conditions in the previous experiments such as the unknown thickness of the inflow boundary layer. Hence, the goal of these measurements is to get data of a turbulent boundary layer at well defined and known boundary conditions like an undisturbed flow in the test section, a defined length between the origin of the boundary layer, i.e., the leading edge, and the measurement location, a hydraulically smooth surface of the flat plat, and no pressure gradient.

2 Experiment

2.1 Wind Tunnel Setup

The investigations were carried out in the $8 \times 6\ m^2$ test section of the DNW-LLF at Reynolds numbers up to $Re_\theta = 54600$. An elliptical contour at the leading edge guaranteed an completely attached flow. The distance from the leading edge to the measurement position on the hydraulically smooth surface of the plate was 13.5 m. The freestream velocity was varied between 40 m/s and 60 m/s yielding a zero pressure gradient turbulent flat plate boundary layer with a maximum thickness of 150 mm in the measurement cross section.

2.2 Measurement-Techniques

Two triple hot-wire probes, developed and manufactured at the Aerodynamisches Institut Aachen, in combination with two TSI IFA 100 constant temperature anemometers were used to investigate the velocity distribution and the two-point correlations in the boundary layer. Both hot-wire probes were calibrated against the freestream velocity measured with pressure transducer mounted in the nozzle exit. The calibration procedure was repeated after each measurement sequence to guarantee the validity of the calibration for the entire set of hot-wire data. Two triple hot-wire probes were used, one of which could be shifted in the vertical direction and the other one could be moved in the normal, spanwise, and streamwise direction such that geometrically a three-dimensional wedge-like zone above the surface of the flat plate could be analyzed. To get as close as possible to the plate and to be able to minimize the distance to the other hot-wire probe, the fixed probe was installed at an angle of 4° between the streamwise probe axis and the surface of the flat plate and the moving probe at an angle of 7°.

Due to our cooperation with the University of Berlin the measured PIV data, which were recorded under exactly the same experimental conditions, could be used to compare turbulent structures based on single-point data sampled at 10 kHz with plane data recorded at 1 Hz. A 2C-PIV and a 3C-PIV setup with four CCD cameras at a resolution of 1280 x 1024 pixel were used to measure the flow field of the complete boundary layer at x=13.5 m downstream of the leading edge and in the near wall layer. The flow is seeded locally with DEHS in the stagnation chamber at a location, which assures that the seeding impinges upon the leading edge. Hence, the flow in the vicinity of the flat plate contains enough particles with a diameter of 1-2 μm for the 2C- and 3C-PIV measurements. A double Nd:YAG pulsed laser systems with an energy of 300 mJ per pulse is used for the light pulses with a period 10 μs and 8 μs at 40 m/s and 60 m/s, respectively. The statistics in each measurement series is based on 5000 pictures, which are captured with a repetition rate of 1 Hz. The cameras were triggered simultaneously and for the 2C- and 3C-PIV the same light sheet was applied.

3 Results

In Fig. 1 the boundary layer profiles for 40 m/s ($Re_\theta = 41833$) and 60 m/s ($Re_\theta = 56400$) are shown. The distributions are scaled with the friction velocity, which was computed using the semi-empirical relation of Fernholz [2]. The law of the wall is plotted with the standard values for the constants κ=0.4 and C=5.1 as a dashed line. It is evident that there is a difference between the theoretical and the experimental distribution. In [3] Oesterlund and Johansson measured the friction velocity with laser oil interferometry in a flat plate boundary layer with zero pressure gradient and showed a change of the constants for the log law at high Reynolds numbers. This distribution of the log law is presented as a solid line in Fig.1. The constants κ=0.36 and C=4.036 represent a close approximation of the findings introduced in [3] κ=0.38 and C=4.1 in the range of the log layer $1000 \leq y^+ \leq 5000$ and as such they confirm the deviation from the classical values at high Reynolds numbers.

The correlation coefficient R_{uu} in the wall-normal direction is determined by the PIV and the hot-wire technique at $y/\delta = 0.1$ and 0.3 at the freestream velocity of 40 m/s. The comparison of R_{uu} in Fig. 2 shows a satisfactory agreement between the PIV and hot-wire distributions corroborating the ability of PIV to be used to determine distributions of the turbulent structures even at a relatively low sampling rate in high Reynolds number flows. The comparison between the R_{uu} data normal to the flat plate for $y/\delta = 0.1$ and $y/\delta = 0.3$ shows the normal turbulence length scale based on the R_{uu} distribution to be nearly independent from the wall distance in the lower part of the boundary layer.

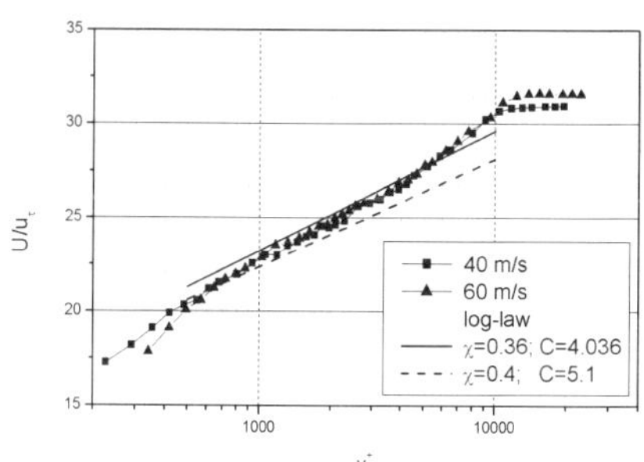

Fig. 1. Boundary layer profile from hot-wire data.

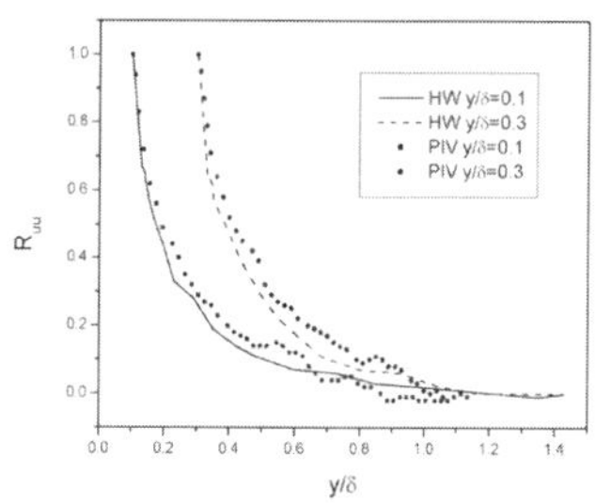

Fig. 2. R_{uu} hot-wire data and PIV data.

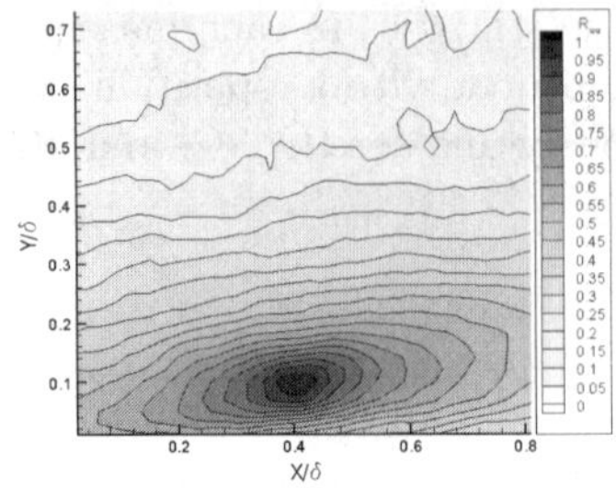

Fig. 3. R_{uu} contours determined by PIV data.

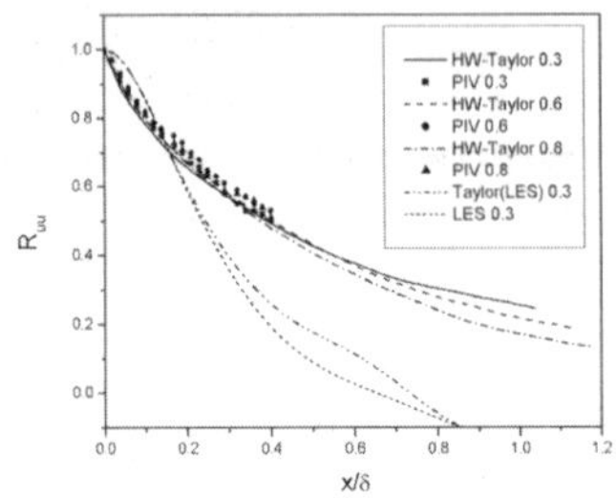

Fig. 4. R_{uu} distribution in streamwise direction.

In Fig. 3 R_{uu} contours at $Re_\theta = 41833$ centered at $x/\delta = 0.4$ and $y/\delta = 0.1$ are shown. From the shape of the isolines and from the orientation of the overall structure it is evident that the turbulent R_{uu} based length scales differ in the streamwise and the wall-normal direction. Furthermore, the maximum scale occurs at a slight inclination angle, approximately $11°$, to the wall surface illustrating the displacement effect of the wall on the turbulent field. Such a displacement is not observed in a free shear layer.

Since the PIV data provide the instantaneous velocity field in the entire x,y-plane these measurements are used to check the Taylor hypothesis. Fig. 4 contains the distribution of the streamwise two-point correlations based on the hot-wire data, i.e., by applying the hypothesis, and on the PIV measurements. The results, which belong to several normal distances of $y/\delta = 0.3$, 0.6, and 0.8 are in excellent agreement, thus the Taylor hypothesis is valid at high Reynolds numbers. Furthermore, the validity of the Taylor hypothesis in the low Reynolds number regime is confirmed by LES data [4] for a flat plate turbulent boundary layer at $Re_\theta = 1410$.

References

1. Barenblatt, G. I.: Scaling laws for fully developed shear flows. Part I: Basic hypotheses and analysis, J. Fluid Mech., vol. 248, pp 513-520,1993.
2. Fernholz, H. H., Krause, E., Nockemann, M., Schober, M.: Comparative measurements in the canoni-cal boundary layer at $Re < 6 \times 10^4$ on the wall of the German-Dutch Windtunnel, Phys. of Fluids, vol. 7, pp. 1275-1281, 1995.
3. Österlund, J. M.,Johansson, A. V.: Measurements in a flat plate turbulant boundary layer, Turbulence shear-flow phenomena 1. TSFP-1, Santa Barbara, California, USA, 2000.
4. El-Askary, W.A., Schröder, W., Meinke,M.: LES of Compressible Wall-Bounded Flows, AIAA Paper 2003-3554, 2003.

Measurements Over a Flat Plate With and Without Suction

Amit Agrawal, Lyazid Djenidi, and R.A. Antonia

Discipline of Mechanical Engineering, University of Newcastle, Australia
Amit.Agrawal@newcastle.edu.au

1 Introduction

Extensive studies have been made to document the structures in turbulent boundary layers. One way to gain additional insights about the structures in the flow is to perturb the boundary layer and examine its response further downstream. We have adopted this approach by applying concentrated suction through a porous suction strip. Suction has also been applied to reduce the drag over airfoils and to control the transition to turbulence. Both experimental [1, 2, 3] and numerical [4] tools have been used by the previous researchers to study suction. Antonia et al. [1] found that the skin friction start increasing just upstream of the suction strip then decreases downstream of the suction region before slowly relaxing to that of an undisturbed flat wall value. Flow was found to be pseudo-laminar with large suction rate downstream of the suction strip. These measurements were confirmed by Oyewala et al. [2] who also investigated the effect of Reynolds number on this flow. Oyewala et al. [2] report that the maxima of turbulent stresses move towards the wall, and there is a decrease in their normalized values. The numerical investigation of [4] showed the pressure distribution with suction.

The purpose of the present study is to extend these earlier works with the view to examine the response of the boundary layer to suction using Laser Doppler Velocimetry (LDV) and Laser Induced Fluorescence (LIF). The experiments were carried out in a closed-circuit vertical water tunnel fabricated using Perspex. The water tunnel is 2 m long with a 250 mm square working section. One of the section walls is used as the testing wall. Suction is applied through a 20 mm porous strip located about 1 m downstream of the leading edge of the wall. More details about the experimental facility can be found in [5]. LDV measurements were undertaken to measure both the streamwise U and wall-normal V velocity components. A continuous 4 W Argon-Ion laser is used for illumination purposes. The Reynolds number R_θ based on momentum thickness θ is $\approx$1000 and suction rate σ [1] is ≈ 3.3 for these measurements.

2 Results and Discussion

2.1 Near Suction Strip Region

In an effort to document the near-field effect of the disturbance caused by suction, measurements were conducted in the vicinity of the suction strip (SS). These measurements also provide the initial condition, and help us understand development of the new layer downstream of the SS. Figure 1 indicates a negative U from the wall up to $y = 1.5$ mm suggesting a backflow in this region. This backflow is because of an adverse streamwise pressure gradient near the trailing edge of the SS [4]. The flow-reversal region is about 1.5 mm in the wall normal direction and 2.1 mm in the streamwise direction. The flow starts to accelerate beyond the flow reversal region due to the deflection of streamlines, which brings the faster moving fluid closer to the wall. Very few studies have reported a negative U downstream of the suction region [3]. This might be because of the small dimensions of the flow reversal region which renders its identification difficult. The suction rate employed here was comparatively large, and our measurements were focused in this very small region making its identification possible.

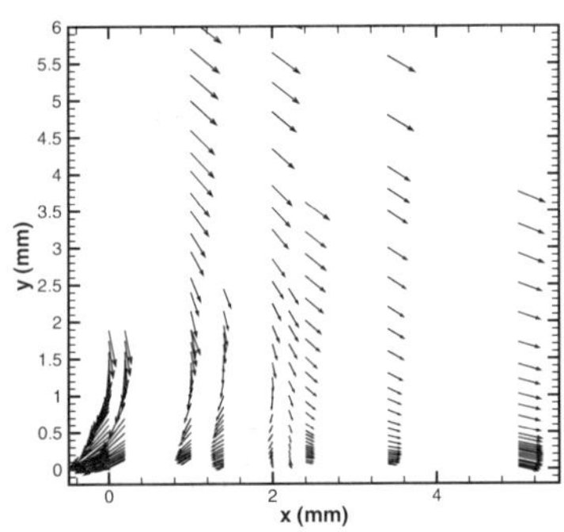

Fig. 1. Vector plot close to the suction strip. The plot is in physical units to give an idea of the region under examination. The origin of the x-axis corresponds to the trailing edge of the suction strip.

Examination of u' (primes denote the fluctuating component) indicates a reduction in rms away from the wall with suction and a large local increase close to wall (i.e., for $x < 1$ mm) (Fig. 2). With suction, the maximum u' is about 2-3 times greater than without it. Therefore, suction deflects the region of maximum turbulence closer to the wall. On the other hand, for $x > 1$ mm, u' and u'/U are substantially reduced with suction. The profiles for v' show the same general trend as u'. However, v'/V actually reduces substantially with suction. The profiles for Reynolds shear stress indicate a reversal in sign just beyond the trailing edge of the suction strip ($x < 1$ mm) (Fig. 2). A possible reason for this is a corresponding reversal in the flow direction. However, the sign of $\overline{u'v'}$ recovers faster than U, i.e., at $x = 1$ mm instead of 2.1 mm. Compared to no suction case, the magnitude of Reynolds shear stress is small for $x > 1$ mm.

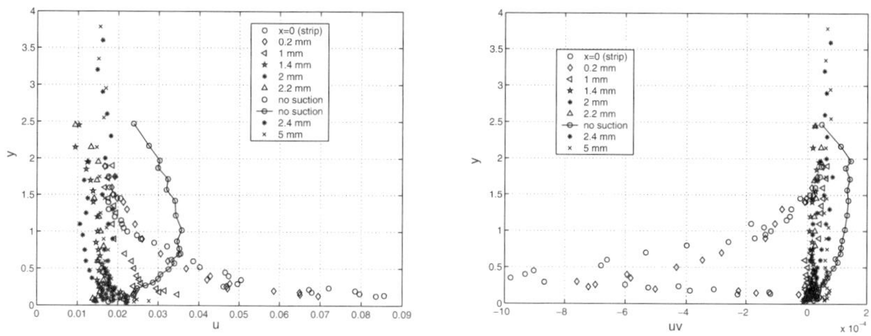

Fig. 2. Streamwise velocity fluctuations *(left)* and Reynolds stress *(right)* close to the suction strip

2.2 Streamwise Variation in Mean and rms Velocities

In this section, we will present results for streamwise variation in velocity at $y+ = 5, 11, 30$, where $y+$ is the wall unit without suction. Figure 3 indicates a large increase of U_s/U_{ns} (subscripts s and ns denotes suction and no suction, respectively) upstream of the SS, which is due to a favorable streamwise pressure gradient generated by the suction. The figure also reveals that the slower moving fluid closer to the wall is more strongly accelerated. The streamwise velocity presents a maximum (minimum) at about the leading (trailing) edge. After the minima, U starts to recover with a slight overshoot, which is somewhat smaller closer to the wall. Finally, U_s/U_{ns} appears to tend to 1.2 indicating that the layer does not return to its non-disturbed state downstream of the strip in agreement with [1, 2, 4]. This is explained by the fact that the disturbed boundary layer is subject to a "new initial condition" as apparent from the previous section. Figure 3 shows little change in V upstream of the strip and a maximum V on the SS, and V_s/V_{ns} tends to 1 downstream of the SS. Due to the small value of V_{ns} there is some experimental jitter on the profiles.

Profiles for u' show a decrease with suction with a minimum just downstream of the trailing edge of the SS. The recovery involves a slight overshoot, and the final value appears to asymptote to 1. Acceleration due to suction and relaminarization of the flow are attributed as the reasons for the decrease in the rms upstream and downstream of the SS. The profiles for v' is somewhat similar to u'.

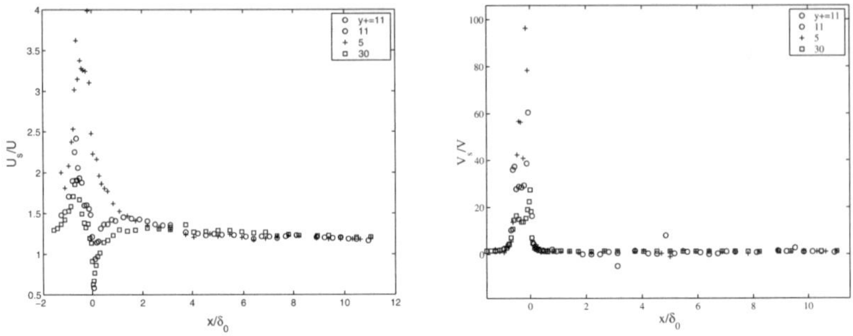

Fig. 3. Streamwise variation in mean streamwise *(left)* and wall-normal *(right)* velocities. δ_0 is the boundary layer thickness without suction at the leading edge of the strip.

2.3 LIF Image Analysis

LIF measurements, to visualize the low-speed streaks near the wall, were accomplished by introducing a fluorescein dye into the flow trough a transverse slit, and illuminating a plane of the test section with the laser. Results for several suction rates were analyzed. The general observation was that suction stabilizes the streaks, i.e., reduces their oscillation and makes them longer. In other words, the tendency to lift-up and break is reduced as the rate of suction increases, and the streaks remain coherent for a longer distance (or time). These results were confirmed by a statistical analysis. The reduction in lifting of streaks is partly because the hairpin structures responsible for the generation of streaks have been removed. Suction also creates a favorable pressure gradient which inhibits the natural tendency of eruption of these streaks.

3 Conclusions

Measurements in the boundary layer over a flat plate with and without suction revealed a region of backflow close to the suction strip. Strong gradients in mean and rms velocities are generated and the turbulent structures are substantially affected due to suction. These measurements are consistent with those of [1, 2, 3, 4] but have revealed more details about the flow.

References

1. Antonia RA, Zhu Y, Sokolov M (1995) Phys. Fluids 7: 2465-2475
2. Oyewola O, Djenidi, L, Antonia RA (2003) Expt. in Fluids 35: 199-206
3. Yoda M, Westerweel J (2001) Expt. in Fluids 30: 239-245
4. Djenidi, L, Antonia RA (2001) Int. J. Heat Fluid Flow 22: 487-494
5. Djenidi, L, Elavarasan, R, Antonia, RA (1999) J. Fluid Mech. 395: 271-294

MHD Taylor-Couette Flow for Small Magnetic Prandtl Number and With Hall Effect

Rainer Arlt and Günther Rüdiger

Astrophysikalisches Institut Potsdam, An der Sternwarte 16, D-14482 Potsdam, Germany. rarlt@aip.de, gruediger@aip.de

1 The Magnetorotational Instability

Rotating shear flows are very common in astrophysics. Rotation profiles of stars, accretion disks, and galaxies are shear flows. The Rayleigh criterion for stability of a given rotation profile requires an increasing specific angular momentum with distance from the rotation axis. This criterion is fulfilled in nearly all astrophysical objects. The rotation profile in accretion disks obeys roughly $\Omega \sim r^{-3/2}$, where r is the axis distance.

A powerful ingredient to rotating shear flows are magnetic fields, which excite a linear instability even if they are weak in terms of energy compared with the thermal energy. The magnetorotational instability (MRI) has been proven by analytical and numerical studies to be very efficient in generating turbulence. The turbulent flows emerging from the instability lead to outward transport of angular momentum (see e.g. [1], [2], [3]). This is a very promising finding for the problem of the formation of stars.

The MRI had not yet been observed in the laboratory at the time of the Conference. Taylor-Couette (TC) experiments study the flow between two coaxial cylinders with one of them, or both, rotating. If the inner cylinder is rotating – by far the most often studied case in the laboratory – the rotation profile $\Omega = A + B/r^2$ looks similar to the Keplerian one, but is Rayleigh unstable whence not comparable to accretion disks. Nevertheless, the TC flow bears the chance to reproduce the MRI in an experiment.

2 Magnetorotational Instability in a Taylor-Couette Flow

The fact that Chandrasekhar did not report on this MHD instability in [4] is due to his approximation of very low magnetic Prandtl numbers, $\mathrm{Pm} = \nu/\eta$, where ν is the kinematic viscosity and η is the magnetic diffusivity. He had good reasons for that, since fluid metals available have $\mathrm{Pm} \approx 10^{-7}$ to 10^{-5}.

The standard TC flow with a rotating inner cylinder and a ratio of inner radius to outer radius, $R_{\rm in}/R_{\rm out} = 0.5$, gets unstable at a Reynolds number of Re = 68. Figure 1 shows the lines of marginal stability in dependence of the imposed magnetic field, normalized to the Hartmann number Ha. If a fluid of Pm = 1 would exist, a weak magnetic field can reduce the critical Reynolds number by about 10%. At Pm = 10, the critical Reynolds number is as small as 23 (left panel). Because such materials do not exist, we have to go another way: instead of reducing the critical Re for the MRI, we may increase the critical Re for the purely hydrodynamic flow.

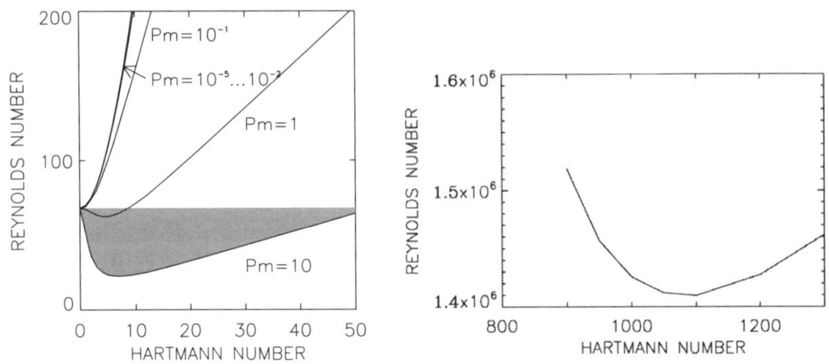

Fig. 1. Lines of marginal stability for the magnetic TC flow with $R_{\rm in}/R_{\rm out} = 0.5$. LEFT: for a resting outer cylinder; RIGHT: for an outer cylinder rotation of 33% of the inner cylinder frequency.

While a rotating inner cylinder alone always means decreasing angular momentum with axis distance, a rotation of the outer cylinder can make the TC flow Rayleigh stable. The change occurs when the outer cylinder rotates at 25% of the frequency of the inner cylinder. The critical Reynolds number of the hydrodynamic case (ordinate crossing in the left panel of Fig. 1) moves to infinity. The critical Reynolds number for the MRI does not move to infinity, as it requires decreasing angular *velocity*. Such a profile is still given. Now we can ask for the minimum of a curve at $\mathrm{Pm} = 10^{-5}$, since the hydrodynamic instability does not exist. The right panel of Fig. 1 shows the line of marginal stability for an outer-cylinder rotation rate of 33%. Although the necessary Reynolds numbers are much higher than for Pm = 1, an experimental configuration for $\mathrm{Re} \sim 10^6$ may be possible with

$$f = 22\,\mathrm{Hz}/(R_{\rm in}/10\,\mathrm{cm})^2 \quad \text{and} \quad B = 1170\,\mathrm{Gauss}/(R_{\rm in}/10\,\mathrm{cm}), \tag{1}$$

where f is the rotation rate of the inner cylinder and B is the magnetic field threading the setup. Although the parameters are not readily achieved in an experiment, especially since the fluid will most likely be liquid sodium, the technical possibilities are about to be sufficient for a magnetic TC experiment studying the MRI.

The success of the experiment is based on the hydrodynamic stability of the setup. Since there is controversy about the experimental stability of a configuration with a rotating outer cylinder, a first experiment must address the stability of the flow which is predicted by linear analysis and observed in [5], but is not found in older experiments [6] and hitherto rarely investigated.

The MHD models also imply electric boundary conditions. The necessary rotation rate for perfectly conducting boundaries is 50% higher than for insulating boundaries. However, the excitation of nonaxisymmetric modes is then possible. Since dynamo-generated magnetic fields must be non-axisymmetric, the magnetic TC flow experiment thus provides the chance of driving a dynamo with a fully "self-determined" flow caused by the MRI. The energy will solely be provided by the walls maintaining their rotation velocity.

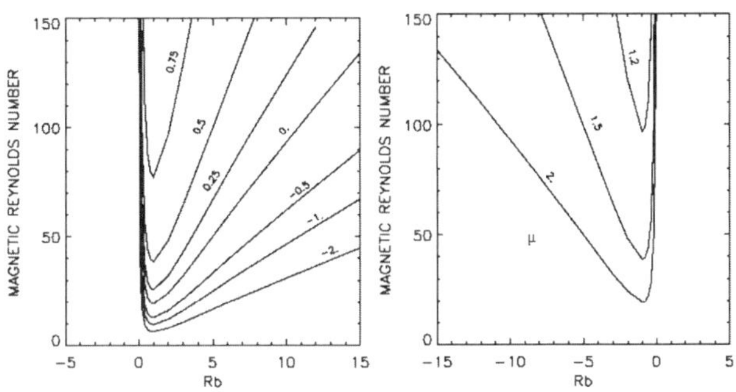

Fig. 2. Lines of marginal stability for the TC flow with Hall effect.

3 Taylor-Couette Flow With Hall Effect

In astrophysical objects, shear flows often occur in thin plasma where the full coupling between ions and electrons may not be given. The magnetohydrodynamic behaviour of the gas is then described by the Hall effect which is an additional electric field due to the independent drift of electrons. The induction equation can be modified to

$$\partial \mathbf{B}/\partial t = \mathrm{curl}\,(\mathbf{u} \times \mathbf{B}) - \mathrm{curl}\,(\mathrm{curl}\,\mathbf{B} \times \mathbf{B}) + \mathrm{Rb}^{-1}\eta \triangle \mathbf{B}\,, \qquad (2)$$

where $\mathbf{u}$ contains the shear flow and Rb is the Hall parameter combining the magnetic background field, the electric conductivity, and the electron number density. Together with the Navier-Stokes equation, we may ask whether the MRI still exists in the Hall regime.

A linear analysis has been carried out on the TC flow [7]. The magnetic Hall-shear instability is indeed found for a certain range of imposed magnetic fields. Figure 2 shows the lines of marginal stability for various rotation profiles.

The left panel shows the cases where the outer cylinder rotates slower or even in the opposite direction than the inner cylinder. The gradient $\mathrm{d}\Omega/\mathrm{d}r$ is negative. The right panel shows the cases of higher outer rotation rate, whence $\mathrm{d}\Omega/\mathrm{d}r > 0$. The astonishing result is that the emergence of an instability now depends on the direction of the magnetic field – parallel or antiparallel to $\boldsymbol{\Omega}$.

The finding is of importance in astrophysics, since concerns have been expressed as to whether or not the MRI is excited in thin gases where the coupling between particles may not be sufficient. The analysis with Hall effect supports the MRI in very thin gas, although only for half of the possible alignments of rotation axis and magnetic fields.

4 A Hall-Dynamo?

The last considerations regard the excitation of a kinematic dynamo from the induction equation with Hall term. They are kinematic in the sense that the flow is given by the laminar TC flow providing shear to the induction equation, but the back-reaction of the magnetic fields is not considered, the Navier-Stokes equation is not solved. The results will only be valid as long as the magnetic energy is small compared to the kinetic energy. A one-dimensional domain extends in the z-direction and is subject to a vertical magnetic field B_z, normalized to the Hall parameter Rb. The linearized equations for such a Hall dynamo read

$$\frac{\partial B_x}{\partial t} = \frac{\partial^2 B_x}{\partial z^2} + \mathrm{Rb}\frac{\partial^2 B_y}{\partial z^2} \quad \text{and} \quad \frac{\partial B_y}{\partial t} = \frac{\partial^2 B_y}{\partial z^2} + C_\Omega B_x - \mathrm{Rb}\frac{\partial^2 B_x}{\partial z^2}, \tag{3}$$

where $C_\Omega = (\mathrm{d}u_y/\mathrm{d}x)R/\eta$ measures the shear in x-direction. A vacuum exterior is assumed, whence $B_x(\pm 1) = B_y(\pm 1) = 0$ as boundary conditions.

Growing modes for the magnetic fields are found for a minimum shear of $C_\Omega > \pi^2/2$. At any stronger shear, a certain range of imposed magnetic fields leads to growing solutions. Recent computations in two dimensions and with galactic rotation profiles confirm the dynamo instability [8]. Future simulations in 3D must show whether the mechanism serves as a real dynamo which can amplify magnetic fields from arbitrarily small initial fields.

References

1. Balbus SA, Hawley JF (1991) Astrophys. J. 376:214
2. Brandenburg A, Nordlund Å, Stein RF, Torkelsson U (1995) Astrophys. J. 446:741
3. Arlt R, Rüdiger G (2001) Astron. Astrophys. 374:1035
4. Chandrasekhar S (1960) Proc. Nat. Acad. Sci., 46:53
5. Schultz-Grunow F (1959) Z. Angew. Math. Mech. 39:101
6. Wendt (1933) Ing. Arch. 4:577
7. Rüdiger G, Shalybkov D (2004), Phys. Rev. E 69a:6303
8. Elstner D, personal communications, January 2004

Laser-Cantilever-Anemometer

S. Barth[1], H. Koch[1], J. Peinke[1], J. Burgold[2], and H. Wurmus[2]

[1] Carl von Ossietzky University, 26111 Oldenburg, Germany
[2] Technical University of Ilmenau, 98684 Ilmenau, Germany

Summary. Based on Patent application DE 198 22 125.8-52, we present a technical description of a new temporal and spatial high resolving anemometer for gas and liquid flows. The measurement principle is based on the technique of an atomic force microscope where microstructured cantilevers are used to detect extreme small forces. We show the sensor as a small compact unit and present first measurements and characterizations.

Principles of Function

Based on the technique of atomic force microscopes (AFM) [1] we use microstructured cantilevers as sensitive element with a typical length $l = 160\mu m$, a width $w = 30\mu m$ and a height $h = 1 - 3\mu m$ normally used to resolve atomic distances and forces. To perform local velocity measurements in a fluid we use the deflection s of the cantilever due to the drag force as the basic signal. Different to the normal application in atomic force microscopes we have surface loads from the fluid instead of loads acting only on the tip of the cantilever. The deflection s of the tip of the cantilever is therefore described by

$$s = \frac{l^3}{8} \frac{F}{E \cdot I_a} = \frac{3 \cdot l^4 \cdot c_d \cdot \frac{\rho}{2} \cdot v^2 \cdot w}{2 \cdot E \cdot w \cdot h^3} = \frac{3 \cdot c_d \cdot \rho \cdot v^2 \cdot l^4}{4 \cdot E \cdot h^3}. \quad (1)$$

with the elasticity modulus E, force F, drag coefficient c_d, fluid density ρ, velocity v and the geometrical moment of inertia I_a which becomes $I_a = \frac{1}{12}hw^3$ for a plate. Note that the deflection is independent of the width of the cantilever. The value of the fluid velocity v is averaged over the cantilever.

To detect the deflection of the cantilever we focus a laser beam on the tip of the cantilever and measure the position of the reflex spot with a psd element (position sensitive detector), see figure 1.

The linear psd element of length L_X which we use in this case to get the position of the incident light yields two electric currents I_{X1} and I_{X2} at the ends of the photosensitive surface (figure 2)

$$I_{X1} = I_0 \frac{L_X - X_B}{L_X}, \qquad I_{X2} = I_0 \frac{X_B}{L_X}, \tag{2}$$

with I_0 as the light intensity and L_X, L_B as different lengths [2]. The following quantity

$$\frac{I_{X1} - I_{X2}}{I_{X1} + I_{X2}} = \frac{I_0 \left(\frac{L_X - X_B}{L_X} - \frac{X_B}{L_X} \right)}{I_0 \left(\frac{L_X - X_B}{L_X} + \frac{X_B}{L_X} \right)} = 1 - \frac{2}{L_X} X_B. \tag{3}$$

leads to an expression which is a linear function of the reflex position and independent of the intensity I_0. Any changes in the light intensity which are acting on the whole psd element will not disturb the position measurement. Thus our measurement is insensitive to the ambient light or variances in the laser intensity.

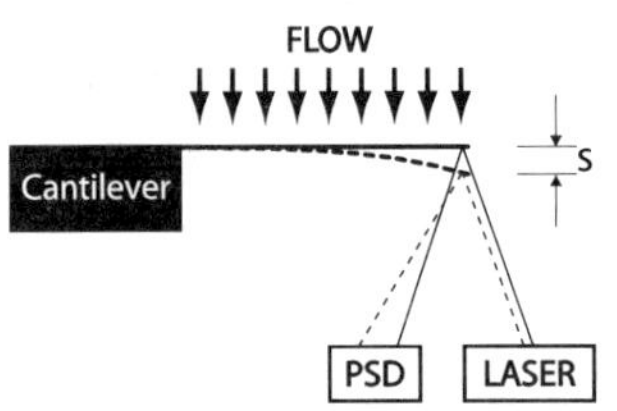

Fig. 1. Cantilever deflection s with flow (dashed lines) and without flow (solid lines). Detected by a laser and a position sensitive detector (PSD).

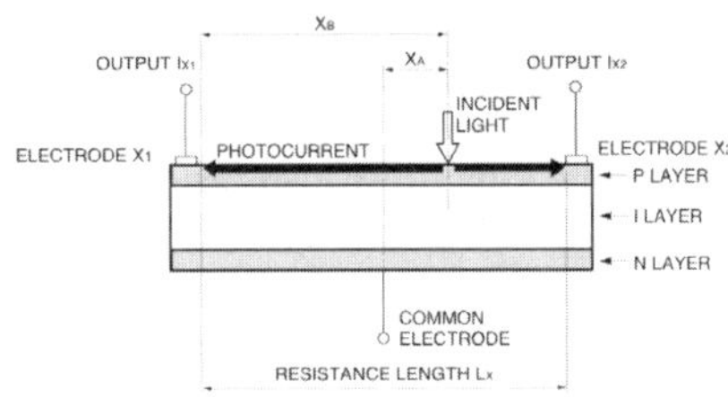

Fig. 2. Schematic illustration of the psd element with the used nomenclature.

Construction

The construction of the LCA is shown in figure 3. At the tip of the sensor a small arm is screwed. At its tip the cantilever is glued, see figure 4. A waterproof enclosure contains the psd element with its amplifier electronics, the laser with optical components and step motors to adjust the laser orientation.

To measure the velocity with high accuracy the chip which is the support of the cantilever is designed as shown in figure 4. Etching a window into the chip gives the advantage that the fluid can pass through the chip itself. The cantilever is hold in position by two $20 \mu m$ thick bridges. Thus the disturbance of the flow around the cantilever is kept small.

The arm on which the cantilever chip is glued has a face surface that is not bigger as the cantilever chip itself, see figure 4. The arm becomes thinner to its top in three steps to reduce the dimensions as much as possible but still keeping the mechanical stability.

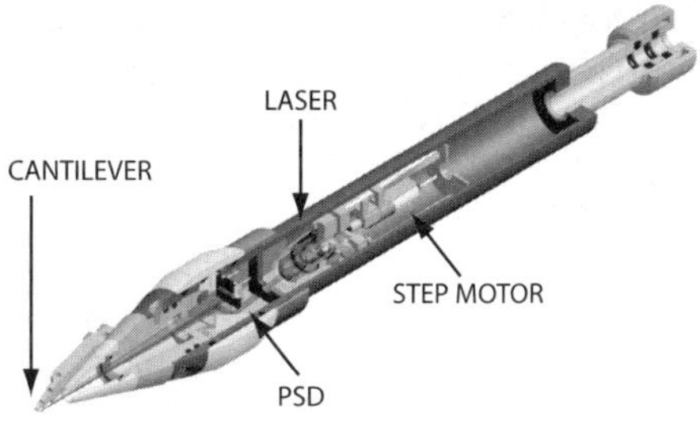

Fig. 3. 3D illustration of the LCA.

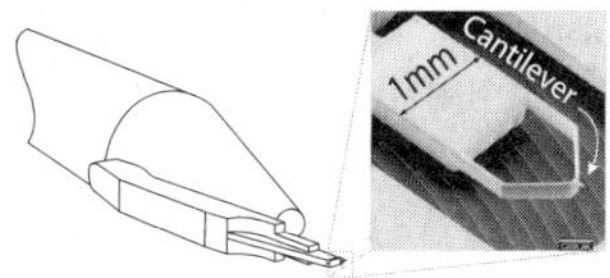

Fig. 4. Top view of the LCA and the arm on which tip the cantilever is glued. In the REM picture a cantilever chip is shown. The cantilever is the rectangular at the tip of the triangle.

First Measurements and Results

In this section first measurements and their results are compared with those from a StreamLine CTA system with a hotwire probe (Dantec 55P01). In fig. 5 the calibration curve of the LCA is shown. Further measurements are based on a freejet experiment with air into air: nozzle-diameter $d = 8mm$, velocity at the nozzle exit $v = 32.8\frac{m}{s}$, $Re = 17500$. Turbulence measurements at a distance of $37.5d$ from the nozzle in the centre of the freejet with $\bar{v} = 12.34\frac{m}{s}$ were performed.

Regarding the power spectrum of the air freejet (figure 6) one can see that the LCA and the hotwire measurements coincide within the interial range of $-\frac{5}{3}$ (Kolmogorov law) and down to the dissipation range ($f < 10^4 Hz$). Note that there are a lot of interfering signals in the hotwire signal at high frequencies (starting at $10^4 Hz$) which don't appear in the cantilever signal because of the shielded enclosure (inset in figure 6).

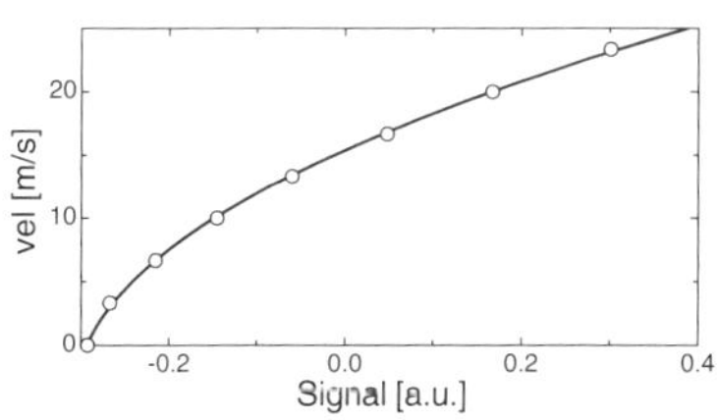

Fig. 5. LCA calibration in the described freejet experiment, performed directly in the laminar flow of the nozzle exit.

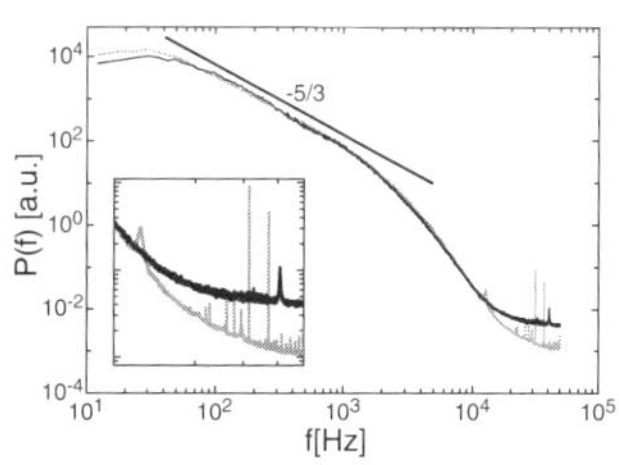

Fig. 6. Power spectrum of the freejet experiment, cantilever: black line, hotwire: gray line.

Next we investigate the multiscaling properties of the turbulent signals. This represents an advanced more detailed analysis. The multiscale analysis corresponds to general two point correlation analysis. We calculated the probability distribution functions (pdfs) for different length scales of velocity increments $v(r) = v(x) - v(x + r)$ and their higher order structurfunctions $\langle \delta v(r)^n \rangle$. To convert the time series in spatial structures we use Taylor's hypothesis. As one can see in figure 7 the pdfs coincide quite nicely. Note that knowing the pdf we also know the $\langle \delta v(r)^n \rangle$ moments.

A further characterization is done by using extended self-similarity (ESS) [3], $\langle \delta v(r)^n \rangle \propto \langle |\delta v(r)^3| \rangle^{\xi_n}$. In figure 8 the scaling exponent ξ_n of structure functions of degree n is plotted. The LCA yields the same results as the hotwire even for $n > 8$ where the estimation of the scaling exponent normally becomes very uncertain.

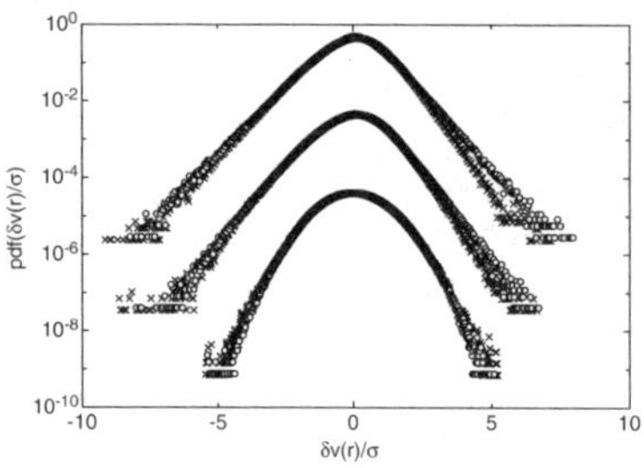

Fig. 7. Velocity increment pdfs of the hotwire (crosses) and the LCA (circles) for the length scales $1.97mm$, $4.69mm$ and $106.23mm$ (from top to bottom)

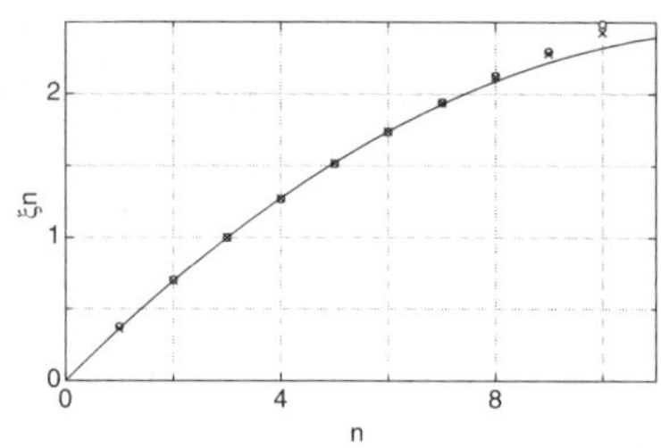

Fig. 8. Scaling exponent ξ_n estimated by extended-self similarity from the LCA dataset (circles) and the hotwire dataset (crosses). The solid line represents a fir with Kolmogorov 1961.

Conclusion

A new type of anemometer based on microstructured cantilevers was presented. The principles, construction and first measurements were shown. The first results were compared with a hotwire anemometer and the quality of the results is comparable. The next steps will be investigations in water experiments and further improvements in the technique.

References

1. G. Binning, C.F. Quate and Ch. Gerber Atomic Force Microscope *Phys. Rev. Lett., 56(9)*, 1986.
2. Jacob Fraden AIP Handbook of modern Sensors. Physics, Designs and Applications *Springer*, 1993.
3. R. Benzi, S. Ciliberto, R. Tripiccione, C. Baudet & S. Succi Extended self-similarity in turbulent flows *Phys. Rev. E 48(1), 29*, 1993.

Heteroclinic Cycles of Type II in the (2,3) Interaction in the GEOFLOW-Experiment

P. Beltrame and C. Egbers

Dept. of Aerodynamics and Fluid Mechanics (LAS), Brandenburg Technical University (BTU), Siemens-Halske-Ring 1, D-03044 Cottbus, Germany.
beltrame@tu-cottbus.de

1 Introduction

Microgravity experiments of thermal convection in a spherical shell under a central force field are important to understand of large scale astro- and geophysical motions such in the outer core of the Earth. In the GEOFLOW-experiment a thermal convection is produced by heating the inner sphere and cooling the outer one. A central force field, similar to the one acting on planets, is simulated using the dielectrophoretic effect [1]. To turn off the unidirectional gravitation under terrestrial conditions, it is undertaken on the International Space Station (ISS).
Beyond a critical Rayleigh number R_a, i.e. a critical temperature difference, the basic state ("pure" conduction) becomes unstable. Generically one spherical mode ℓ is unstable, than only stationary or traveling waves solutions are expected. But for a critical aspect ratio η (the ratio of the inner to the outer radius of the shell) two consecutive modes $(\ell, \ell+1)$ interact. Due to the spherical symmetry, the bifurcation diagram is very rich (time-dependent dynamics, intermittencies...) and in particular, heteroclinic cycles connecting equilibria occur. These motions are very reminiscent of the aperiodic reversal of the Earth's magnetic field in geological times.
The (1,2) interaction was thoroughly studied: first, in a numerical point of view by Friedrich and Haken [2], but the systematically study of heteroclinic cycles and their robustness was undertaken by Armbruster and Chossat in [3] using group theoretic methods. Later, Chossat and Guyard [4] are given a generic classification of heteroclinic cycles for the $(\ell, \ell+1)$ interactions. Because of experimental requirements the (1,2) interaction is not possible [5] and in the (2,3) one, such heteroclinic cycles can only occur in the neighbourhood of a critical Prandtl number P_{rc}. As in [4], we are not able in [6] to observe numerically the so-called type II heteroclinic cycle. So, we aim to finding such cycles with an open basin of attraction by varying the Prandtl number.

2 Mathematical Background

The governing equations for the perturbed system are the classical system of PDE's of Navier-Stokes in the Boussinesq approximation for the temperature, and with Dirichlet boundaries for the velocity (viscous fluid) and for the temperature perturbation (imposed temperature) [5]. After a suitable nondimensionalization, it appears three parameters for our problem: the aspect ratio η, the Prandtl number P_r and the number λ, which is proportional to the square root of the R_a number [5].
For $(\lambda_c; \eta_c) \simeq (14; 0.33)$, the basic state stability is undoing and the modes 2 and 3 interact, i.e. the eigenspace V is the direct sum of $V_2 \oplus V_3$, where V_ℓ is the $2\ell + 1$-dimensional complexified space of spherical harmonics of order ℓ. We shall denote $z_n^\ell, n \in \{-\ell, ..., \ell\}$ the coordinates of V_ℓ with the "reality" condition $z_{-n}^\ell = (-1)^n \bar{z}_n^\ell$ (see [5] for more precisions).
The original system of PDE's is reduced on its center manifolds, which is parameterized by the space V. The dynamics near the onset can be examined by varying the two system parameters λ and η. So, the bifurcation equations is governed by a system of ODE's which it consists of 5 equations for $z_n^{(2)}$ and 7 equations for $z_n^{(3)}$. For the numerical simulations, we use the Taylor series expansion of the third order [5].
Yet, we focus on the cycles for which the heteroclinic connection between equilibria lies in an invariant plane (cycles "forced" by the $O(3)$ symmetry). Then [4], the equilibria are the both axisymmetric solutions $\alpha_\pm$ (with reversal motions) of mode 2 ($O(2) \oplus \mathbf{Z}_c^2$ isotropy group). The invariant planes are the fixed-point spaces corresponding to the subgroup of $O(2) \oplus \mathbf{Z}_c^2$. The table 1 gives the three planes which can lead to a such heteroclinic cycle. Contrary to the planes P_2 and P_3, the transition in P_1 contains two other copies $\alpha_\pm^2$, $\alpha_\pm^3$ of $\alpha_\pm$ deduced of it by the rotations of $2\pi/3$ and $-2\pi/3$. In the physical space, there are the three axisymmetric motions of the three orthogonal axis. So, the transition connects two axisymmetric steady-states with different axis. A cycle with a such transition is called of type II.
In order to obtain two equilibria α (pitchfork bifurcation), it is necessary that the pure quadratic term a of the mode 2 is close to zero, this condition is satisfied for $P_{rc} \simeq 0.24$. We have not observed this cycle in [6] for two reasons: the existence domain of the type II is included in the type I and the contracting eigenvalue of α in the plane P_1 is very small compared to the others (it is proportional to a). So the cycle of type II is destroyed by the other one. Because a has to be small, we search a Prandtl number range for which the existence domain of type II is not included in the type I.

3 Results

The stability analysis of the equilibria α in the plane $P_i, i \in \{1, 2, 3\}$ shows that the existence region of the type II cycle can be decomposed as $R_2 \cup R_3$,

Table 1. Partial isotropy lattice and corresponding fixed-point subspaces.

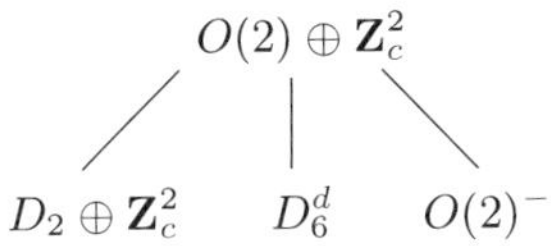

Σ	$Fix(\Sigma)$
$O(2) \oplus \mathbf{Z}_c^2$	$D = \{z_0^2\}$
$D_2 \oplus \mathbf{Z}_c^2$	$P_1 = \{z_0^2, Re z_2^2\}$
D_6^d	$P_2 = \{z_0^2\} \oplus \{Im z_3^3\}$
$O(2)^-$	$P_3 = \{z_0^2\} \oplus \{z_0^3\}$

where R_3 is the region of the cycle I, if the Prandtl number is in the range:

$$0.28 < P_r < 0.24 \simeq P_{rc}. \tag{1}$$

In this case, the transition planes for the cycle I are P_2 and P_3 and for the cycle II P_1 and P_2 (Tab. 1). If $P_r \simeq 0.28$, the region R_2 vanishes, and if $P_r \simeq P_{rc}$, the coefficient a is zero. So, the choice of P_r is a compromise between the attractivity of the cycle ($a \neq 0$) and the size of R_2. We have given an example for $P_r = 0.25$. The width with respect the λ direction of the region R_2 is very thin: 0.1% (Fig. 1.) An example of cycle is showed in the figure 2. The transition duration is very short compared to the stays near the equilibria (Fig. 2-b). Because the contracting eigenvalue in P_1 is small, the transition in the plane P_2 is faster than the one in P_1. In agreement with the theoretic results of [4], the cycle is robust against small perturbations. The numerical simulation shows an important basin of attraction: for various initial conditions, after some transients, the cycle is established. We are not able to prove the orbital asymptotic stability with the criterion described in [4]. However the strict inequality (8.1) of criterion of [4] is fast satisfied: it is an equality in our case.
In the region 3, the cycles I and II can occur. We obtain intermittencies, which are looking like the cycle II: connections between axisymmetric solutions of different axis. This competition between cycles is the subject of a new study.

Acknowledgments

Fundings from the ESA Postdoc are greatfully acknowledged.

References

1. Egbers Ch *et al.* (2003) Adv. Space Res. 32(2):171–180
2. Friedrich R, Haken H (1986) The American Physical Society 34(3):2100–2120
3. Armbruster D, Chossat P (1991) PhysicaD 50:155-176
4. Chossat P, Guyard F (1996) J. Nonlinear Sci. 6:201–2385-176
5. Beltrame Ph, Egbers Ch, Hollerbach R (2003) Adv. Space Res. 32(2):191–197
6. Beltrame Ph . *et al.* (2003) Heteroclinic Cycles in the GEOFLOW-Experiment on the ISS In: Marqués F (eds) Nonlinear Dynamics in Fluids. Barcelona.

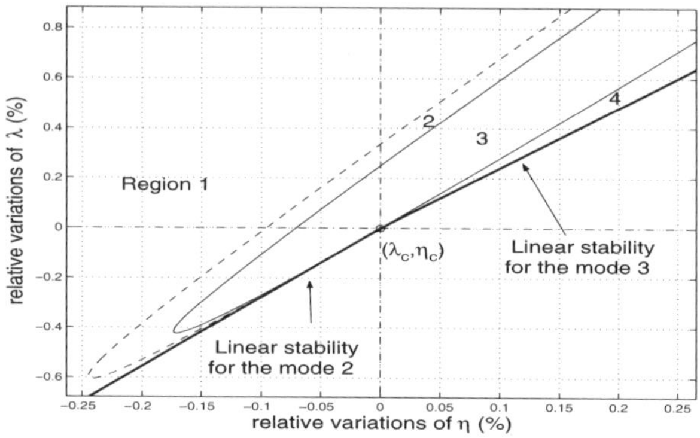

Fig. 1. Regions of heteroclinic cycles in the (λ, η) parameters plane..

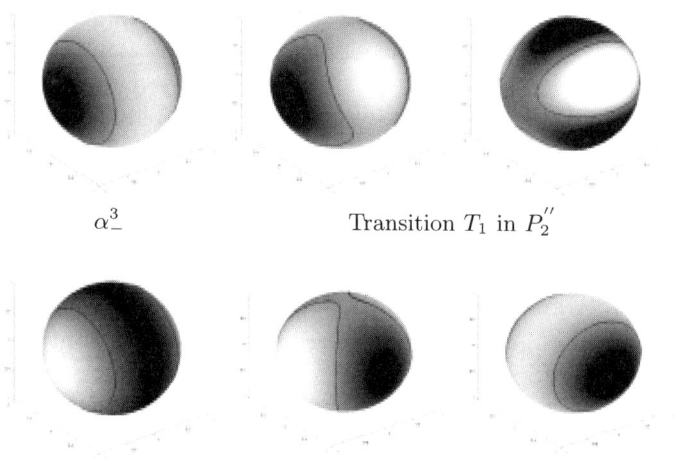

α^3_- Transition T_1 in $P_2^{''}$

α^3_+ Transition T_2 in P_1 α^2_-

(a) Radial velocity: the up-(light) and the down-welling (dark) motion.

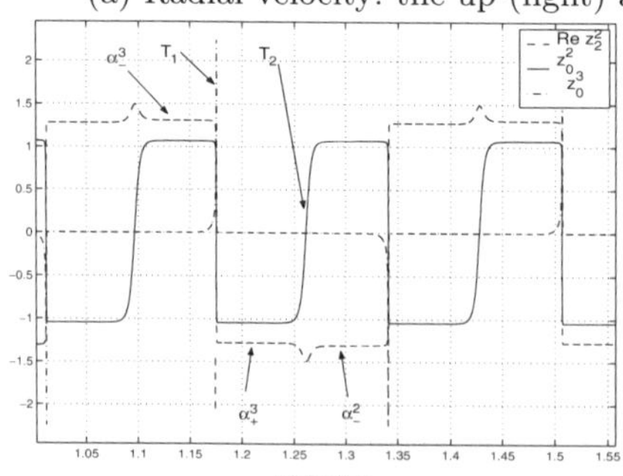

(b) Time variations of the amplitudes

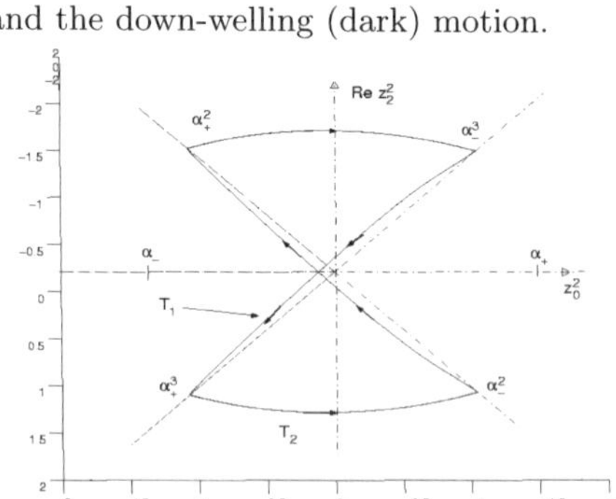

(c) Phase portrait

Fig. 2. A heteroclinic cycle of type II for $(\tilde{\lambda}, \tilde{\eta}) = (0.65\%, 0.1\%)$

Experimental Visualization of Streamwise Streaks in the Boundary Layers of Rayleigh–Bénard Convection

Tomi Haramina and Andreas Tilgner

Geophysical Institute, University of Göttingen, Herzberger Landstrasse 180, 37075 Göttingen, Germany tomi.haramina@geo.physik.uni-goettingen.de

Summary. Two series of visualization experiments were performed in the boundary layer of Rayleigh–Bénard convection with the goal to get more information on the behavior and characteristics of streaks elongated in the streamwise direction observed in the boundary layers. These streamwise streaks are a new type of coherent structures in the Rayleigh–Bénard convection.

1 Introduction

In the study of turbulent flow much attention is paid to understanding "coherent structures" in the boundary layers. Coherent structures are defined as distinctive patterns of motion (in contrast to turbulent, unordered flow) which occur repeatedly, during a limited period of time. The motivation for their investigation lies in the fact that their understanding helps in modeling and controlling turbulent flows as well as in shedding light on the dynamics of the flows [1]. The best known type of coherent structure in thermal convection is the thermal plume (e.g. [2]). Swirls and waves living in the interface between the boundary layer and the bulk have also been visualized [3]. In the present work we present a new type of coherent structure in the boundary layer of thermal convection. If the plate of the convection box is covered with a thin layer of dye, the dye accumulates in streaks aligned with the large scale mean flow. Qualitative (Sect. 3) and quantitative (Sect. 4) characteristics of the streaks are presented together with an explanation of the observed statistics of the streak separation. Section 2 explains the visualization method used.

2 Experimental Setup and Method

The experiments were performed in a cubic cell 20 cm in side and filled with water. The sidewalls are made of 10 mm thick plexyglas whereas the top and the bottom plates are 10 mm thick silver-coated plates. The upper plate is cooled by water circulating through pipes welded on the plate, and the lower plate is heated by electrical heaters attached to it. The temperatures of the plates are kept uniform in space and time to better then 0.1 K.

Flow is visualized using the "pH method" [4]: a pH indicator (in our case "thymol blue") is added to water and solution is titrated with HCl and $NaOH$ just to the acid side of the end point of the indicator. Then an electrical potential difference is applied between the upper and lower plates, such that the lower plate is used as cathode. Because of hydrolysis there is an increase in pH in the vicinity of the cathode, which causes a change in color of the pH indicator in that area (thymol blue changes color from orange to dark blue). If a dyed fluid particle is carried away from the electrode, it looses through diffusion its pH value and thus also the contrast to the surrounding fluid. In that way only a narrow area in the very vicinity of the cathode (lower plate) can be colored (depending on the voltage applied), making the method highly convenient for visualization in the boundary layers.

3 Some Qualitative Characteristics of the Streaks

The first set of experiments was performed so that the amount of dye in the cell was gradually increasing. In that way it was possible to observe the behavior of the flow in different parts of the cell in streamwise direction, as well as at different distances from the plate. Figures 1–3 show three representative examples of that set of experiments. The figures show the lower plate of the cell. The mean large scale flow is from the lower left to the upper right corner of the picture. The regular grid drawn on the plate is used for distance calibration. On Fig. 1 a very thin layer near the lower plate was dyed, and in the two subsequent figures the amount of dye is increased. The streaks are aligned with the mean flow, they first appear at about half the way from one corner to the other (Fig. 1) and they are advected by the mean flow. A streak is an indicator of flow convergence. This requires upwelling of the fluid above the streaks and downwelling in between them. If the streamlines of upwelling and downwelling flow are closed is not yet clear. The pattern of closed streamlines producing a single streak would be a pair of counterrotating vortices superposed on, and aligned with, the mean flow. This is a well known pattern in isothermal shear flows, but in our case the Reynolds number of the boundary layer is only about $Re = 30$, which is far below values typically found in the flows where the streamwise vortices were observed. This indicates that the mechanism of the streak production is not the same. No streaks perpendicular to the mean flow are ever observed in the downstream part of the flow. Figure 2 reveals some details of the upstream part of the flow and further away from the plate (the dye layer is thicker). Here, circular patterns, or parts of it, can be identified. They originate from downward plumes which hit the lower plate and push the fluid outwards. This is the mechanism of wave production described in detail and visualized using another technique in [3]. Also, the lifting of the streaks can be observed in the downstream part of the flow. In the visualization shown in Fig. 3 more dye is produced than in the one in Fig 2. The circular patterns are here even better seen than in Fig. 2. The upward flow (upward part of the large scale mean flow) in the downstream corner is detectable as well.

Fig. 1. A few streamwise streaks are marked by arrows

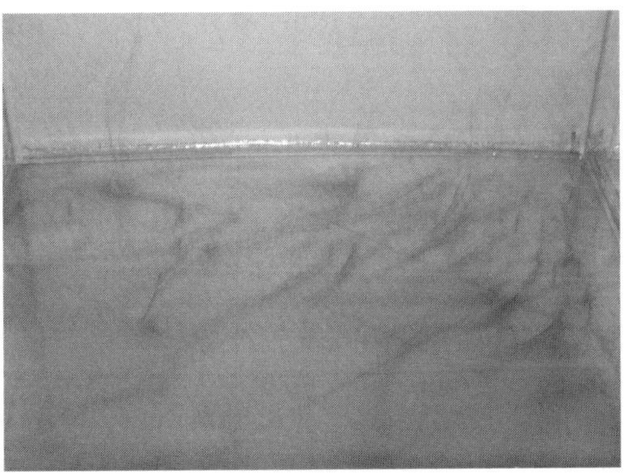

Fig. 2. More dye then in Fig. 1: circular structures are identifiable in the upstream part of the flow

4 Streak Separation

In the second set of experiments photographs similar to the one in Fig. 1 were analysed for different Rayleigh and Prandtl numbers (Table 1, first three columns). Photographs are taken using a 5 Megapixel CCD camera every 60 seconds so that the pictures are statistically independent. The position of each streak along the lines AB and CD (see Fig. 1) was determined manually. AB is the main diagonal perpendicular to the mean flow and CD is parallel to the AB and translated $2\sqrt{2}$ cm downstream.

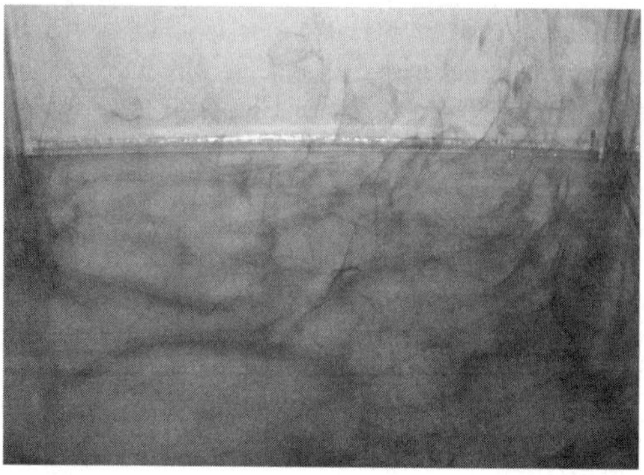

Fig. 3. Even more dye is added to the flow. Upwelling can be detected downstream

Table 1. Minimal streak separation δ determined for different Ra and Pr according to procedure described in the text

Ra	Pr	Position (see Fig. 1)	δ (cm)
5.0×10^8	6.5	CD	1.0
1.3×10^9	6.7	CD	0.6
1.3×10^9	6.7	AB	0.8
1.4×10^9	3.6	CD	0.8
2.0×10^9	6.2	CD	1.0

The apparent spanwise periodicity in the statistical distribution of lateral separation of the streaks (peak in the histogram in Fig. 4) could imply that they result from a linear instability of the flow (combination of shear flow and Rayleigh–Taylor instabilities).On the other hand, one can assume that the streaks appear mutually independently, with the only constraint that they may not appear closer to each other than their lateral size (otherwise they merge). It can be shown that under these assumptions the histogram of streak separations $H(D)$ constructed based on N_p photographs, with bin size ΔD and with N streaks per photograph on average, and the probability $P(n,L)$ to find n streaks at any given time in an interval of length L are given by [5]:

$$H(D) = \begin{cases} \frac{L-D-\delta}{L-\delta} p e^{-p(D-\delta)} \Delta D N_p N & , \delta < D < L-\delta \\ 0 & , \text{otherwise} \end{cases} \tag{1}$$

$$\begin{aligned} P(0,L) &= \begin{cases} 1 & , L < \delta \\ e^{-p(L-\delta)} & , L \geq \delta \end{cases} \\ P(n,L) &= \begin{cases} 0 & , L < n\delta \\ \int_0^{L-\delta} p e^{-px} P(n-1, L-x-\delta)\, dx & , L \geq n\delta \end{cases} \end{aligned} \tag{2}$$

where L denotes the total length of the segment (in our case diagonals AB or CD), δ is the lateral size of the streaks and $p\mathrm{d}x$ is the probability to find a streak in an interval $\mathrm{d}x$.

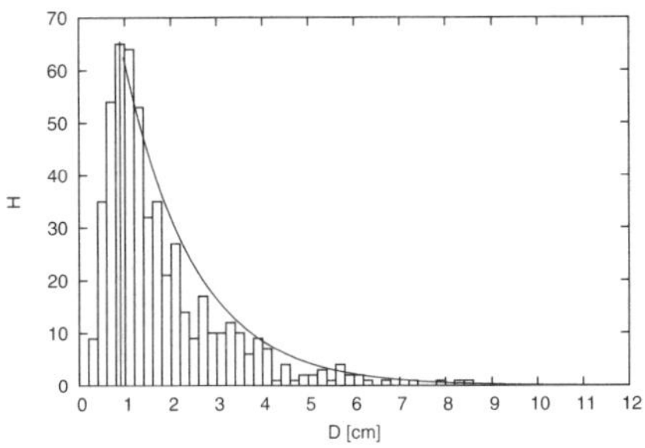

Fig. 4. Streak separation D measured along the line denoted as CD on Fig. 1 (*bar chart*) and theoretical curve given by (1) for $\Delta D = 0.2$ mm, $\delta = 0.8$ cm and p = 0.6

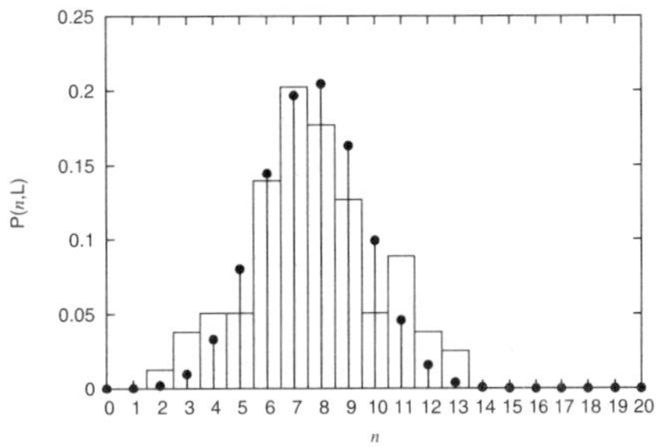

Fig. 5. The bar chart is the probability distribution for the number of streaks n visible in any one photograph for the same data as in Fig. 4. The dots are $P(n, L)$ from (2)

The assumption that δ is equal for all structures is the reason that $H(D)$ equals zero for $D < \delta$. In order to better reproduce the observed data in the region $D < \delta$, a more complex model using a distribution of δ instead of its constant value could be employed. On the other hand, that procedure would introduce more adjustable parameters, but would not contribute much to illuminating the phenomenon. That is why we take for δ the D at which the experimental histogram has its maximum and adjust p so that $P(n, L)$ fits as well as possible the experimental histogram for the relative number of streaks in any moment. Then we use this p in the expression (1) for $H(D)$ and verify that it acceptably fits the experimental histogram for the distribution of streak separation. Figures 4 and 5 show the results of this procedure together with the experimental histograms for $Ra = 1.4 \times 10^9$ and $Pr = 3.6$. Apart from limitation of fixed δ, it is seen that the experimental data are satisfactorily fitted by (1) and (2), that is, under the assumption of mutual independence of the streaks.

Values for δ determined as the distance at which the separation distribution has the peak are given in Table 1 for different values of Rayleigh and Prandtl numbers. The experimental histograms are constructed with bin size 2 mm, which slightly influences the determination of δ, nevertheless the important point here is that for the modest range of parameters in our experiments there is no evident tendency in change of δ with the parameters.

5 Conclusion

A not yet described type of coherent structure — streamwise streaks — is visualized in the boundary layer of Rayleigh–Bénard convection. Two sets of experiments are performed: (1) by controlling the amount of dye in the flow some qualitative characteristics of the streaks are observed, (2) the positions of streaks are recorded and statistical analysis of their separation is performed. Experiments are done for several values of Ra and Pr (Table 1) in a cell of the aspect ratio 1. Finally, a theoretical explanation for the observed distribution of the streak separation is proposed.

The streaks start to appear at about half the way from one corner of the cell to the other, they are aligned with the large scale mean flow and appear to have a preferred separation. This is suggested also by the statistical analysis — the distribution of the streak separations has a pronounced peak. We explained the distribution by assuming that the streaks appear at random and mutually independently. The theoretical curves obtained that way fit the experimental data satisfactorily.

Further experiments in a wider range of parameters (Ra, Pr, geometry) are necessary to answer the remaining question: can linear stability analysis predict the preferred streak separation; if yes, what is its scaling with the parameters of the flow? Furthermore, it remained unclear if the streaks are produced by pairs of counterrotating vortices, or whether the flow upwells above the streaks and downwells over the broader area between the streaks. In order to answer that question, another experimental technique or/and numerical simulation should be employed.

References

1. Robinson S (1991) Annu Rev Fluid Mech 23:601–639
2. Taylor G (1950) Proc R Soc London A 201:175
3. Zocchi G, Moses E, Libchaber A (1990) Physica A 166:387–407
4. Baker D (1966) J Fluid Mech 26:573–575
5. Haramina T, Tilgner A (2004) submitted

Spatial Correlations in Turbulent Shear Flows

A. Jachens[1], B. Eckhardt[1], K. Knobloch[2], and H.-H. Fernholz[2]

[1] Fachbereich Physik, Philipps Universität Marburg, 35032 Maburg, Germany
arne.jachens@physik.uni-marburg.de,
bruno.eckhardt@physik.uni-marburg.de

[2] Hermann-Föttinger-Institut für Strömungsmechanik, Technische Universität Berlin, 10623 Berlin, Germany knobloch@pi.tu-berlin.de

1 Introduction

The short range correlations, both in time and in space, carry information about small scale coherent structures and their dynamics. Various kinds of vortices, streaks or waves have been identified [1, 2] and considerable efforts have gone into identifying their dynamical origins and evolution. In boundary layers the non-normal amplification or lift-up effect [3, 4, 5, 6, 7, 8, 9] is often an important source for coherent structures. The dynamics of the coherent structures can be identified in temporal cross correlation functions between vertical and streamwise velocity components [10], as can be easily seen from a noise-driven linear system [11]. We will here analyze spatial correlation functions in data from numerical simulations and PIV data from a wind tunnel in order to gain further information about the typical size of these coherent structures.

2 Correlators

We consider numerical simualtions of turbulent flows where a suitable body force sustains a mean gradient [12]. The simulations use a pseudo-spectral code with $128 \times 64 \times 128$ nodes and a spatial domain $[2\pi \times 1 \times 2\pi]$. The system is averaged over a time interval of length $t = 20/S$, where S is the mean shear. Such long averaging times can only be achieved with the body force driving that maintains a mean shear. The Taylor-Reynolds number is about $Re_\lambda \sim 45$.

The experimental data are obtained from Particle Image Velocimetry (PIV) measurements in the Large Low-speed Facility of the German Dutch Windtunnel (DNW-LLF). The measurements were taken towards the end of a 15 m long test section, at a flow speed of 40 m/s. The boundary layer thickness δ_{99} is about 160 mm. For further information about the experiment and the flow, see [13, 14] and the contributions to these proceedings.

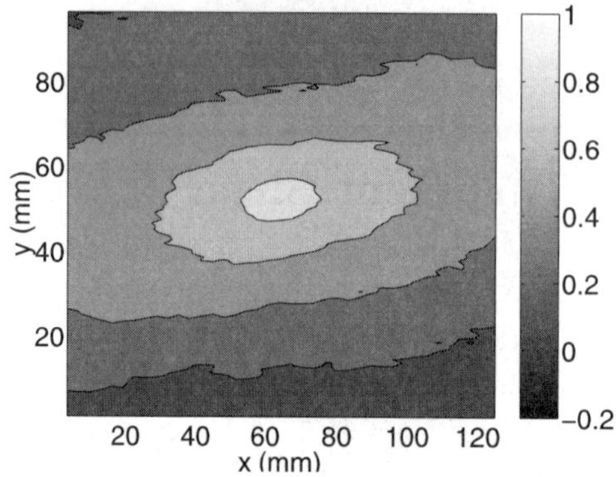

Fig. 1. Correlation function R_{uu} from measurements in the DNW. y_P is $52.5\,mm$, corresponding to $y_P/\delta_{99} = 0.33$.

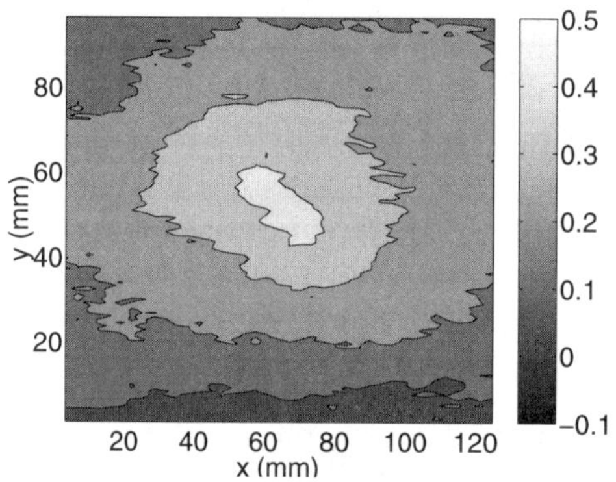

Fig. 2. Cross correlation R_{vu} from measurements in the DNW for the same y_P an in Fig. 1.

The main object of the present study is the behaviour of the spatial autocorrelation function of the streamwise velocity,

$$R_{uu}(\Delta x, y|y_P) = \langle u(\Delta x, y)u(0, y_P)\rangle \,, \tag{1}$$

where y_P is the vertical position of the probe. The experimental data are averaged over 2000 realizations. The numerical data are averaged over horizontal planes and 100 time frames.

The correlation function is symmetric under the point symmetry $(x, y) \rightarrow (-x, -y)$. The contour lines of the correlation function are ellipsoidal, with the main axis not aligned with the coordinate axis, i.e., neither the streamwise direction nor the one for the shear. The angle varies between $11° - 13°$; variations with Reynolds number seem to be weak.

The flow in the numerical simulations is not a boundary layer proper, but a flow driven with a mean shear and bounded by two free-slip surfaces. Since the profile has an inflection in the middle, the comparison should relate the thickness of the boundary layer in the experiment δ_{99} to half the gap width in the numerical simulation, $L_y/2$.

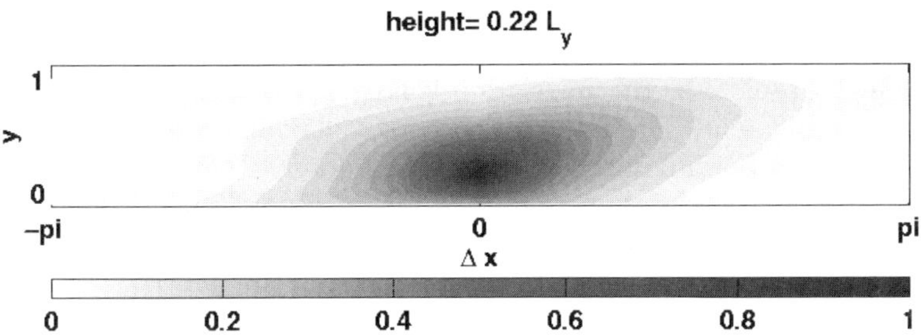

Fig. 3. Normalized correlation function R_{uu} from DNS.

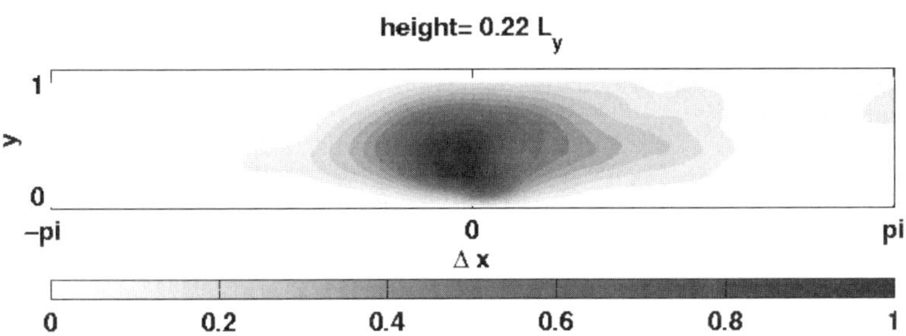

Fig. 4. Normalized cross correlation R_{vu} from DNS.

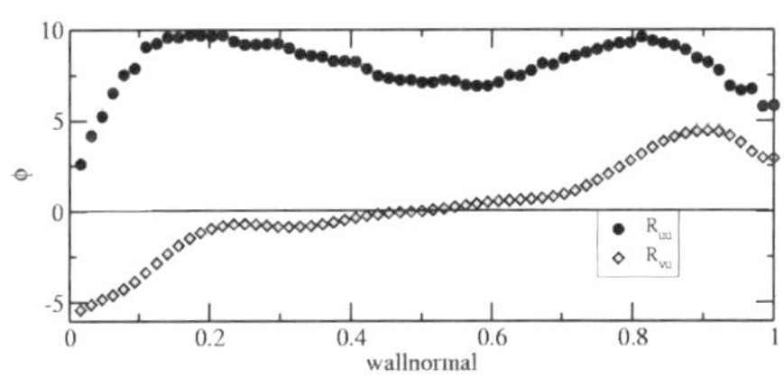

Fig. 5. Tilt angle of R_{uu} (dots) and of R_{vu} (diamonds) as a function of the position between the plates in the DNS data. The angle was determined from an analysis of the second moments for the contours at a given value.

3 Cross Correlations

Temporal cross correlations between vertical and streamwise velocity components are proved to be useful in an analysis of the non-normal amplification [10]. Their spatial counterpart can be expected to carry information about the size of coherent regions responsible for vertical momentum transport. The appropriate quantity is the cross-correlation between the vertical and streamwise velocity component,

$$R_{vu}(\Delta x, y|y_P) = \langle v(\Delta x, y)u(0, y_P)\rangle \,, \tag{2}$$

obtained by a similar averaging procedure as above. Note that while R_{uu} has a point symmetry, R_{vu} does not. The contours of the correlation function are again ellipsoidal as the correlation lengths in streamwise and normal direction are different. The full widths at half maximum in R_{vu} and R_{uu} in streamwise and normal direction compare favorably. This shows that the coherent structures that dominate the vertical transport (as measured by R_{vu}) are of the same size as the streaks in the streamwise velocity. But in contrast to the auto-correlation function the contours of R_{vu} are not tilted with respect to the axes: moving across the boundary layer there does not seem to be a significant tilt, certainly not as strong as in R_{uu} (see Fig. 5).

4 Summary

The small scale correlations between streamwise and vertical velocities in turbulent shear flows continue to provide insights into the small scale structures and their dynamics. The decay of correlations gives an idea of the size of the significant structures. Perhaps the most puzzling observation is the tilt in orientation of the autocorrelation function for the streamwise velocity, which seems to have an upward angle of about 10° with a weak Reynolds-number dependence at best.

We are grateful to thank the Deutsche Forschungsgemeinschaft, the Neumann Center for Computing at the Forschungszentrum Jülich and the Deutsch-Niederländische Windkanal for support.

References

1. Townsend AA (1976) The structure of turbulent shear flows, 2nd ed. Cambridge University Press, Cambridge
2. Holmes P, Lumley JL, and Berkooz G (1996), Turbulence, coherent structures, dynamical systems and symmetry, Cambridge University Press, Cambridge
3. Landahl M (1980) J Fluid Mech **98**, 243
4. Boberg L and Brosa U (1988) Z Naturforsch A **43**, 697
5. Trefethen, N.L., Trefethen, A., Reddy, S.C. and Driscoll, T.A. (1992) Science **261**, 578
6. Grossmann S (2000) Rev Mod Phys **72**, 603
7. Hamilton JM, Kim J, and Waleffe F (1995) J Fluid Mech **287**, 317
8. Moehlis J, Faisst H and Eckhardt B (2004) New J Phys (submitted)
9. Moehlis J, Faisst H and Eckhardt B (2004) in preparation
10. Eckhardt B, Jachens A and Schumacher J (2004) in Reynolds Number Scaling in Turbulent Flow, Smits AJ (ed) Kluwer, Dordrecht, pp 253-256
11. Eckhardt B and Pandit R (2003) Europ J Phys B **33**, 373
12. Schumacher J and Eckhardt B (2000), Europhys. Lett. **52**, 627
13. Knobloch K and Fernholz HH (2004) in Reynolds Number Scaling in Turbulent Flow, Smits AJ (ed) Kluwer, Dordrecht, pp 11-16
14. Abstiens R, Schröder W and Limberg W (2004) in Reynolds Number Scaling in Turbulent Flow, Smits AJ (ed) Kluwer, Dordrecht, pp 279-284

Hot-Wire and PIV Measurements in a High-Reynolds Number Turbulent Boundary Layer

K. Knobloch and H.-H. Fernholz

Hermann-Föttinger-Institut für Strömungsmechanik
TU Berlin, Str. d. 17.Juni135, 10623 Berlin knobloch@pi.tu-berlin.de

A canonical turbulent boundary layer on a large flat plate has been investigated for Reynolds numbers $4 \times 10^4 \leq Re_{\delta 2} \leq 5.5 \times 10^4$. This experimental investigation includes hot-wire measurements of the mean and fluctuating velocity profiles and of two-point velocity correlations in the wall-normal direction. Complementary Particle Image Velocimetry (PIV) measurements in a plane given by the streamwise and the wall-normal direction provide additional information about the instantaneous velocity field.

The comparison between these two measuring techniques is of special importance since we are not aware of published PIV measurements at comparable Reynolds numbers. In addition, the validity of Taylor's hypothesis towards the high Reynolds-number range was investigated.

The project is part of the "High Reynolds Number Turbulence" group of the "Initiative of Turbulence" and also provides experimental data for other projects.

1 Experimental Setup

The experiments were performed in the low-disturbance environment of the German-Dutch windtunnel (DNW-LLF) using measuring equipment of the boundary layer group at the Hermann-Föttinger-Institut. The test runs had free-stream velocities of approximately 40 and 60 m / s.

The boundary layer investigated developed along a large smooth-surface flat plate (roughness 1.2 μm) with the measuring location about 13.5 m downstream of the ellipitical leading edge. The boundary layer thickness δ_{99} is about 160 mm at this postion, the Reynolds numbers (based on momentum thickness) Re_{δ_2} are approximately 39200 and 54300 for the two test cases.

For the hot-wire measurements standard normal-wire and triple-wire miniature probes were used which were calibrated in situ. The traverse gear could move two probes independently of each other in the wall-normal direction in order to obtain space correlations of the three fluctuating velocity components.

The laser light sheet for the PIV measurements was aligned with the mean flow (x-) direction and the wall-normal (y-) direction. Four PCO cameras, two of them capturing the in-plane velocity components over the whole boundary layer height, and two capturing the inner-most region in a three-component stereo setup, were used. Each camera had 1280×1024 pixel and 12 bit resolution. For the evaluation of turbulence quantities 5000 pictures were taken with every camera for each test case.

2 Mean and Turbulence Quantities

Profiles of the mean velocity are plotted in inner-law scaling in Fig. 1. The three sets agree very closely with each other and are – over a range of almost two decades – well represented by the log law. The smaller range for the PIV data is due to the chosen optical magnification which yields a higher resolution in the inner part for the stereo setup.

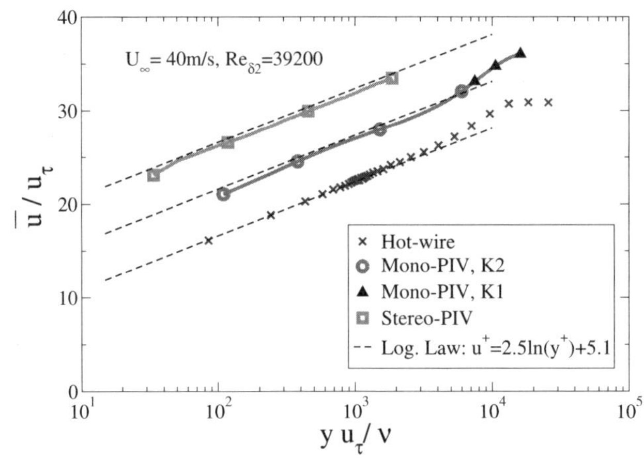

Fig. 1. Mean velocity profiles: PIV data sets are plotted with a constant offset and only a few symbols for clarity

The rms values of the streamwise and wall-normal fluctuations u' and v' are shown in Fig. 2 and Fig. 3, respectively. For the streamwise component, there is reasonable agreement between hot-wire and PIV data. For the wall-normal component, two reference data sets – one from hot-wire measurements at a lower Reynolds number and one from LDV measurements [1] at a comparable Reynolds number – are plotted. Hot-wire and LDV data – despite the different Reynolds numbers – compare well in the log-law region while the PIV data set is substantially lower. A closer examination of the distribution functions of the PIV data (not shown here) reveals the influence of the discrete nature of the data aquisition - often referred to as peak-locking. The small dynamic

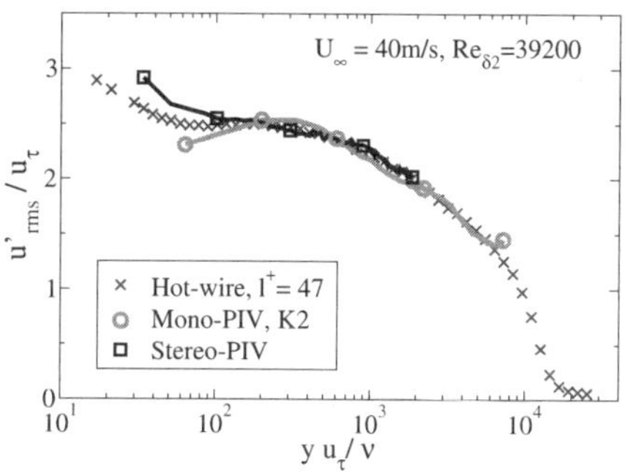

Fig. 2. Streamwise component u'$_{rms}$

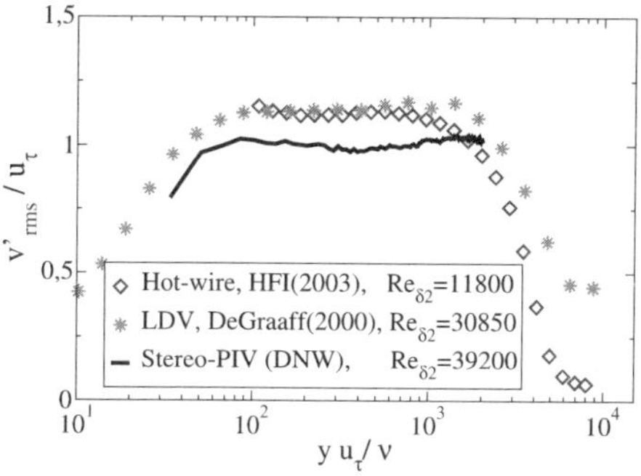

Fig. 3. Wall-normal component v'$_{rms}$

range and a mean value of zero for the wall-normal component contribute to this underestimate of v'$_{rms}$. Improved algorithms for the estimation of the displacement at the sub-pixel level are expected to improve the accuracy and are currently under investigation.

3 Velocity Correlations

Even if the magnitude of the Reynolds stresses for the PIV measurements is not yet considered to be very accurate, the influence on the *normalized* correlation functions is almost negligible (see e.g. the measurments of Grant [2]). Two examples for the streamwise and the wall-normal component are shown in Figs. 4 and 5 where e.g. R_{uu}, the normalized correlation function for the u-component, is given by

$$R_{uu} = \overline{u'(\mathbf{x},t)\, u'(\mathbf{x}+\Delta\mathbf{x}, t+\Delta t)}\,/\,(u'_{rms}(\mathbf{x})\, u'_{rms}(\mathbf{x}+\Delta\mathbf{x}))\,.$$

The different length scales between the two functions are clearly visible as is the anisotropy of these functions which depends strongly on the wall distance

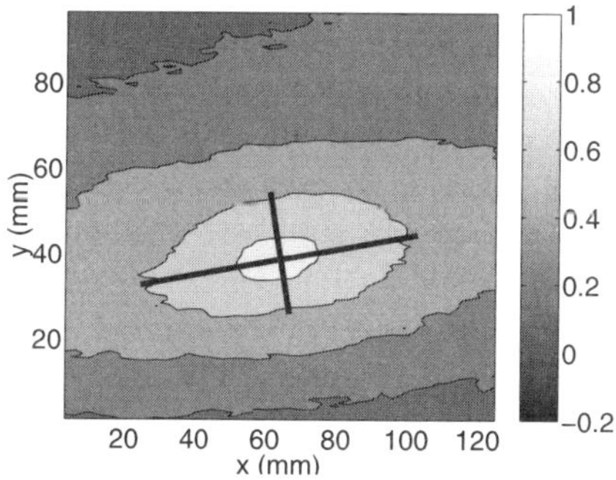

Fig. 4. Normalized correlation function $R_{uu}(y \approx \delta/4)$, $U_\infty = 40\,\mathrm{m/s}$

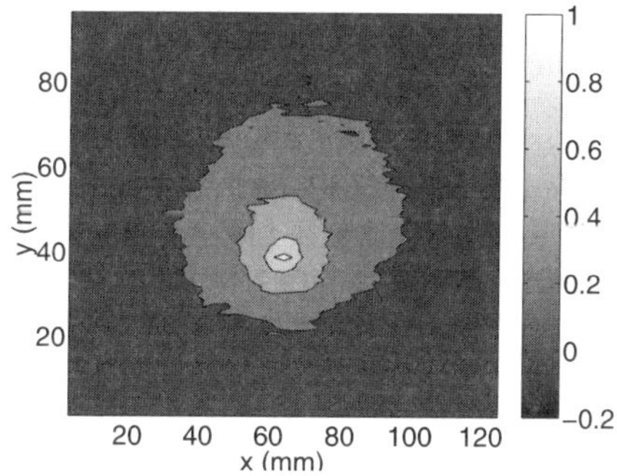

Fig. 5. Normalized correlation function R_{vv} for the same position

(not shown here). The inclination of this structure with respect to the wall suggests the use of different length scales (indicated by the black lines in Fig. 4) than the commonly used scales Λ_x (parallel to the wall) and Λ_y (normal to the wall). A comparision of hot-wire and PIV data is shown in Fig. 6 and 7.

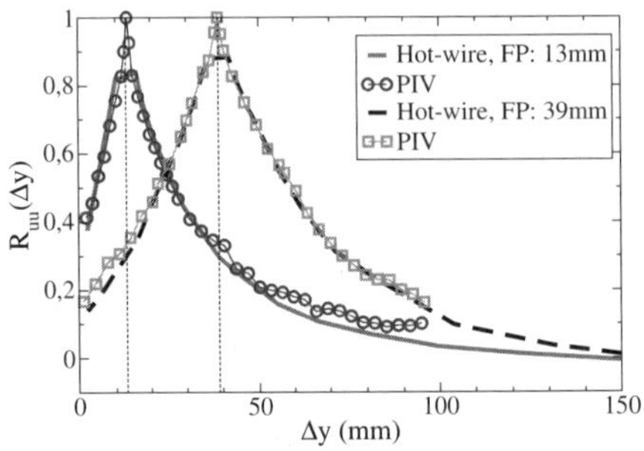

Fig. 6. R_{uu} in wall-normal direction $U_\infty = 40\,m/s$, $Re_{\delta 2} = 39200$

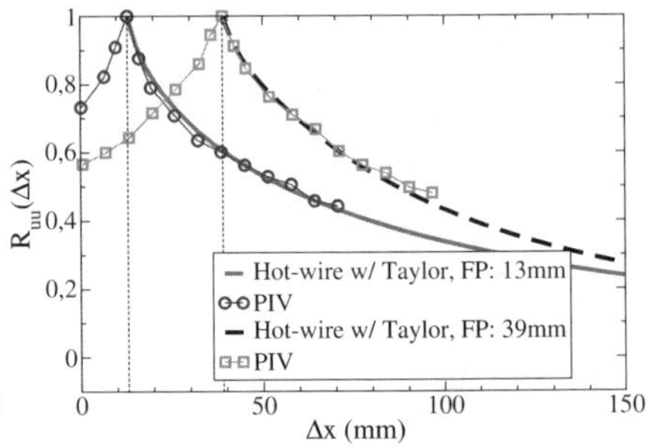

Fig. 7. R_{uu} in streamwise direction, data sets are shifted by 13 mm and 39 mm, respectively for clarity

While the correlation function for the wall-normal direction can be examined directly, the hot-wire measurements do not allow a streamwise separation. Instead, the plotted curve was calculated from the temporal autocorrelation function using Taylor's hypothesis with the local mean velocity as the convection velocity. The collapse of the curves proves the validity of the hypothesis also in this Reynolds-number range.

4 Conclusions

The comparison of hot-wire and PIV measurements for a turbulent boundary layer at $Re_{\delta 2} = 39200$ showed very good agreement for the mean velocity and the streamwise velocity fluctuations. The spanwise component was underestimated by the PIV due to the dynamic resolution of the data, but advanced algorithms may improve the accuracy. The correlation functions were found to be in good agreement and confirmed Taylor's hypothesis for our test.

References

1. DeGraaff DB and Eaton JK (2000) *J Fluid Mech* 422:319–346.
2. Grant HL (1958) *J Fluid Mech* 4:149–190.

Dynamics of Baroclinic Instabilities Using Methods of Nonlinear Time Series Analysis

Th. von Larcher and C. Egbers

Department of Aerodynamics and Fluid Mechanics, Brandenburg University of Technology Cottbus, Germany, P.O. Box 10 13 44, 03013 Cottbus, Germany
larcher@las.tu-cottbus.de

1 Motivation

We study experiments of baroclinic instabilities in a different heated rotating annulus. Characteristic time series are measured with the optical Laser Doppler Velocimetry (LDV) technique. The methods of nonlinear time series analysis are used to investigate the dynamics of baroclinic waves, particulary in transition zones between different flow regimes where complex flows as well as steady waves are observed.

2 Experimental Setup

The rotating annulus has three concentric cylinders with free surfaces (Fig. 1). The experimental gap between the inner and outer chamber is filled with de-ionised water (kin. viscosity ν=1.004 mm^2 s^{-1}). Here, a radial temperature gradient is generated. The inner and outer radius is a=45 mm and b=95 mm and the aspect ratio amounts to 4.4. The radial velocity component of the flow is measured with the fixed LDV thereabout 0.02 m below the surface.

3 Experiments

Figure 2 shows regions of baroclinic waves in the parameter space for water [1]. Different flow regimes are observed depending on the rotation rate and temperature gradient. The Taylor number Ta and thermal Rossby number Ro is defined as:

$$Ta = \frac{4 \times \Omega^2 \times (b-a)^5}{\nu^2 \times d} \qquad\qquad Ro = \frac{g \times d \times \Delta\varrho}{\overline{\varrho} \times \Omega^2 \times (b-a)^2}$$

with d as the depth of the fluid, Ω as the angular velocity, $\overline{\varrho}$ as the average density of the fluid and g as the acceleration due to gravity.

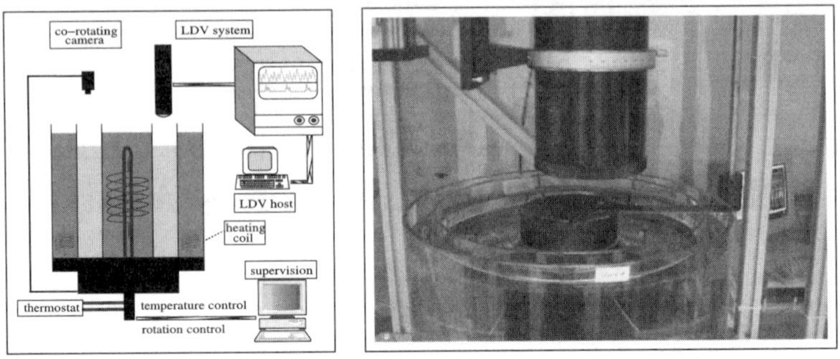

Schematic diagram Side view image

Fig. 1. Experimental setup

Our experiments focus particularly on the first transition zone from axisymmetric basic flow, denoted as upper and lower symmetrical in Fig. 2, to steady baroclinic waves, where low dimensional chaotic flow as well as regular flow occurs (cf. [2], [3], [4]). Here, 'amplitude vacillation' (AV) and 'modulated amplitude vacillation' (MAV) waves are observed, which oscillation frequency is not constant in contrast to AV-waves.

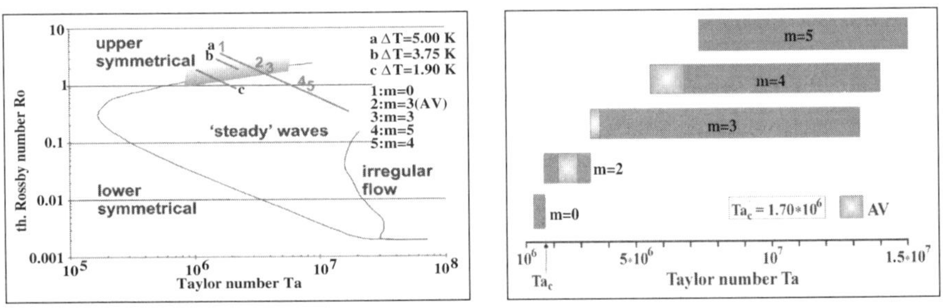

Diagram of flow regimes. Straight lines indicate LDV profiles. Diagram of the parameter space for $\Delta T = 5.0\,\mathrm{K}$

Fig. 2. (**a**) State of the art: parameter space for water [1], m as wave number, first transition zone is shaded (**b**) parameter space of LDV profile $\Delta T = 5.0\,\mathrm{K}$ (see Sect.4)

4 Nonlinear Time Series Analysis

Figure 2 shows the analysis of profile $\Delta T = 5.0\,\mathrm{K}$. Hysteresis effects are found at a wide range of the parameter space. The critical Taylor number Ta_c is 1.7×10^6, stable baroclinic waves with wave number up to 5 as well as AV-waves are observed. The transition to turbulence sets in at $\mathrm{Ta} \approx 1.22 \times 10^7$.

To characterize the nonlinear dynamics of baroclinic instabilities the phase space is reconstructed with the method of time-delayed coordinates [5]. Dynamic variables, e.g. the largest Lyapunov exponent λ_1, the correlation dimension (D_2) and pointwise dimension (D_p) are then calculated to analyse the complexity of the flow. Reccurrence Plots are also calculated.

Figure 3 shows the analysis of an amplitude vacillation wave (m=2, AV). Here, λ_1 is small but significant positive and D_p is calculated to 3.5, whereas the plot of D_2 give no clear result (cf. Fig. 3 f, g). Instead, Fig. 4 shows an analysis of a steady flow state. Here, $\lambda_1 = 0.0$, $D_2 = 1.2$, $D_p = 1.1$. Note also the difference in the autocorrelation function (ACF), the reconstructed attractor (RA) and the recurrence plot (RP). Taking all results into account low dimensional chaotic flow and regular flow respectively is found at these parameter points.

The principle progression of flow regimes from axisymmetric to irregular flow is portrayed by reconstructed attractors in Fig. 5 (from left: axisymmetric flow, flow in first transition zone, steady wave regime). The complexity of the flow at the first transition zone is significant in contrast to the regular flow regime at higher Taylor numbers.

5 Conclusion

The methods of nonlinear time series analysis enable us to characterize the dynamics of baroclinic instabilities. Both, stable, time dependent and complex flows exist in the first transition zone. With increasing Taylor number, the flow becomes regular before the transition to turbulence sets in.

Acknowledgments

This project is funded by the Deutsche Forschungsgemeinschaft (DFG), grant no. EG 100/3-2.

References

1. Fowlis W W, Hide R (1965) J Atmos Sci 22:541–558
2. Früh W–G and Read P L (1997) Phil Trans Roy Soc Lond A 355:101–153
3. Read P L, Bell M J, Johnson D W and Small R M (1992) J Fluid Mech 238:599–632
4. Sitte B, Egbers C (2000) Higher order dynamics of baroclinic waves. In: Pfister G, Egbers C (eds) Physics of Rotating Fluids, Proc. 11th Int. Couette–Taylor Workshop. Lecture Notes in Physics. Springer, Berlin et. al.
5. Takens F (1981) Detecting strange attractors in turbulence. In: Rand D, Young L S (eds) Dynamical Systems and Turbulence. Lecture Notes in Mathematics. Springer, Berlin

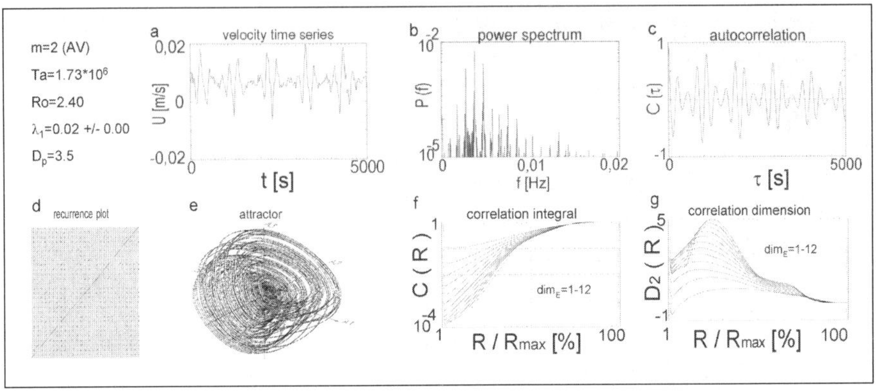

Fig. 3. Nonlinear analysis of an amplitude vacillation wave (m=2, AV): decaying ACF (c), RP shows pointwise structure (d), RA: complex structure (e), non converging plot of D_2, $\dim_E$ as embedding dimension (f, g)

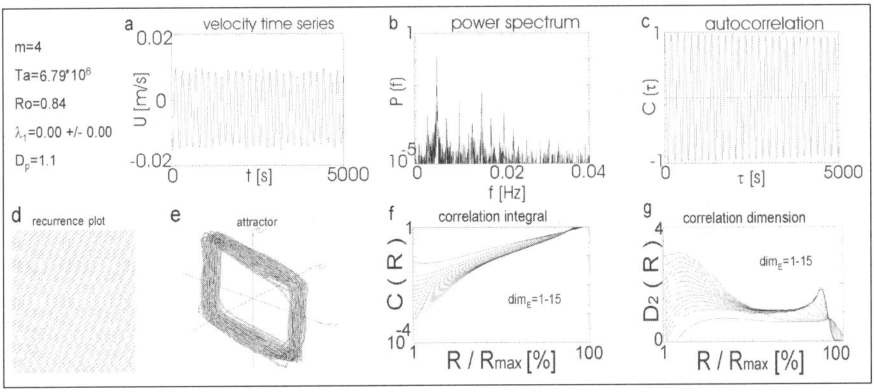

Fig. 4. Nonlinear analysis of a steady baroclinic wave (m=4): periodic ACF (c), RP shows parallel lines (d), RA: torus-like structure (e), converging plot of D_2, $\dim_E$ as embedding dimension (f, g)

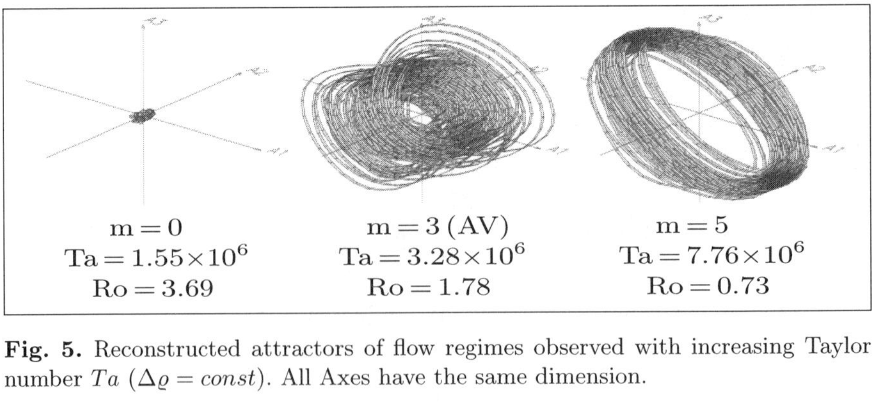

Fig. 5. Reconstructed attractors of flow regimes observed with increasing Taylor number Ta ($\Delta\varrho = const$). All Axes have the same dimension.

Fabrication and Characterization of Miniaturized Thermocouples for Measurements in Flows

M. Munzel and A. Kittel

University of Oldenburg, Institute of Physics, Energy and Semiconductor Research Laboratory Marco.Munzel@uni-oldenburg.de kittel@uni-oldenburg.de

The measurement of thermal fluctuations is important for the investigation of the transport features of passive and active scalars in fluids. As an addition to the established cold-wire technique we present a thermal sensor based on a miniaturized coaxial thermocouple. The advantage of such a sensor is first of all its size. The active area extends only a few hundreds of square nanometers sitting at the tip of a thin glass rod of less than a micrometer in diameter. The preferred field of application of this sensor are all measurement situations which require a high spatial resolution of temperature measurements for example within the boundary layer [1]–[7]. The sensors coaxial setup results from its fabrication as a micropipette and has the advantage of an intrinsic shielding against external distortions. The glass micropipettes contain a core of platinum and are coated with gold and are fabricated similar to the ones in [8]. Because of its chemically inert coating, these sensors are applicable for detecting temperature fluctuations in a large variety of liquids and gases. The fabrication and characterization of these sensors is presented here.

1 Preparation of the Micropipettes

A micropipette puller normally used by microbiologists enables us to fabricate microelectrodes with a metallic core of a few hundred nanometers in diameter insulated by a glass film. These microelectrodes are mainly used in analytical chemistry. A 1 mm thick borosilicate glass tube and a 25 μm platinum wire are smelted and thinned out down to a core size of 200 nm. The achieved tip diameters are 1 μm and less which enhance the spatial and temporal resolution significantly. The heat is supplied via an 15 W cw infrared laser. The result of this fabrication step is an 'open' micropipette as shown in the SEM picture in Fig. 1.

The bar shown on the right hand side of the picture represents a length of 500 nm. It is obvious that the diameter of the free standing Pt-wire sticking

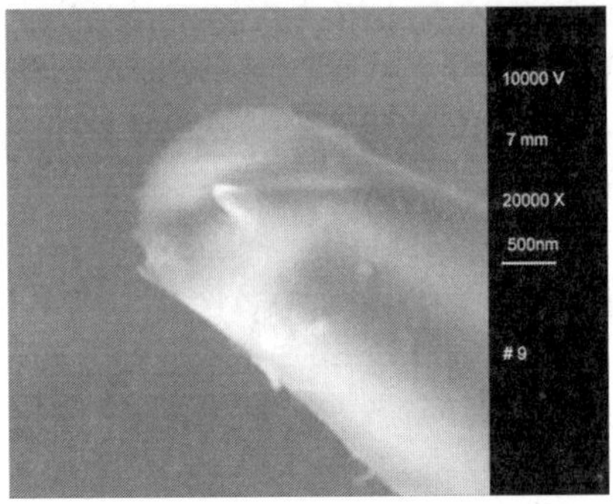

Fig. 1. The electron micrograph shows a free-standing end of a Pt-wire at the tip of a micropipette surrounded by glass. At the left side of the picture the parameters of the electron microscope are quoted.

out of the glass pipette is in the range of 100 nm. The surface of this wire represents the active sensor area after coating the pipette with a 20 nm thick gold film. Before the gold film is evaporated the pipette is glued in a gold plated brass cone with an opening angel of 30°. This design should have a negligible influence on the flow in case of a higher mean velocity like in free-jet experiments if the sensor is pointing against the mean flow. In principle this detail of the design can be easily changed if one is interested in small scale measurements in a boundary layer. Here it would be advantageous to mount the glass capillary directly in the wall of the cell.

2 Design of the Measuring Head

Since we are interested in a as high as possible resolution in temperature and time we have to deal with very small signals at a high bandwidth. To face this challenge we had to bring the preamplifier as close as possible to the sensor in a perfect shielded enclosure. Furthermore this preamplifier has to possess extraordinary noise values. On the other hand an exchange of the sensor should be easily performed. Our design meets these requirements by screwing the cone with the glued sensor together with a tube in which a printed circuit board with the preamplifier is fixed (see Fig. 2). The electrical connections are made by spring loaded, gold plated contacts manufactured by the company Ingun, Germany.

As a preamplifier we used an ultra low noise operational amplifier of the type AD797 fabricated by Analog Device. The bandwidth of the preamplifier is adjusted to the range from DC to 100 kHz, the spectral noise density is below $1.5\,\mathrm{nV}/\sqrt{\mathrm{Hz}}$, and the amplification is 1000. In addition the entire circuit is supplied by batteries to reduce the noise.

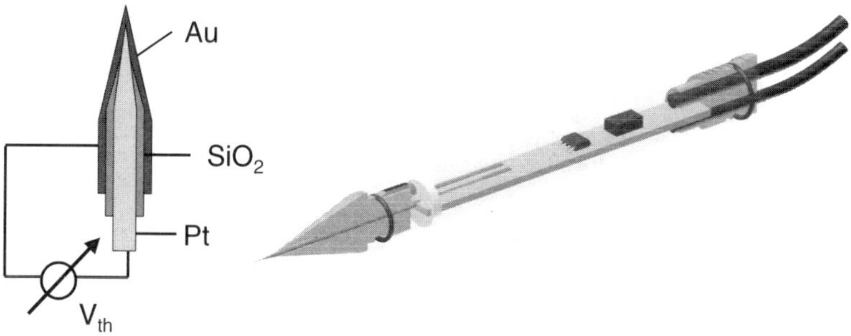

Fig. 2. The basic setup of the thermocouple is shown in the left sketch. The thermocouple is mounted on a tube which serves as a enclosure for the preamplifier as shown on the right.

3 Characterization and First Results

First we calibrated the sensor to determine the temperature from the measured signal. The thermoelectromotive force of a platinum-gold thermocouple was referenced a standard K-type thermocouple in the temperature range of $20 - 100°\mathrm{C}$. This was done using a drop of pump oil sitting inside a coil of resistance wire. On one side the of the drop the sensor was sticking into it and on the other side the commercial sensor. The thermoelectromotive force was increasing almost linearly with temperature and the slope was in accordance with value of $8\,\mu\mathrm{V/K}$ as reported in the literature.

The response time of the thermocouple was measured by means of a chopped and focused 30 mW He-Ne laser. The amplified voltage signal was referenced to the applied light pulse captured by a pin-photo diode. In this case the applied heat pulse has to be redistributed inside the sensor and into the surrounding air. The response time for a $3\,\mu\mathrm{m}$ thermocouple in air was 250 Hz

First measurements with a heated free-jet of water in water are shown in Fig. 3. In the experiment a 2 mm nozzle was reaching from below in a cylindrical water tank of 0.5 m diameter and 1 m height with a free surface. The sensor was located at a distance of 200 mm right above the center of the nozzle outlet pointing against the flow. The speed of the water at the outlet was 4 m/s, 6 m/s, and 9 m/s from light to dark symbols in the diagram of Fig. 3. One can clearly distinguish the inertial and the dissipation region in the spectrum. We observe the extension of the inertial range to higher frequencies in the power spectrum with increasing *Re*-number as described in the relevant text books. At higher frequencies the white noise background can be observed. As expected the response time of the sensor is increased compared to measurements in air. The cut off frequency can not be seen because the signal is already hitting the noise floor.

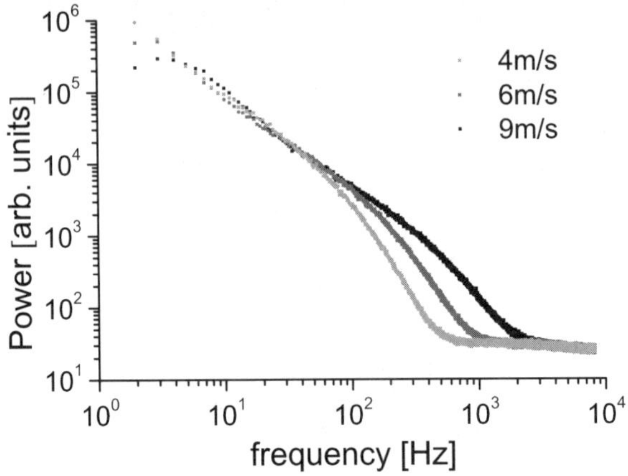

Fig. 3. Power spectrum measured in a heated free-jet (water in water).

To conclude we have demonstrated that a miniaturized thermocouple can be fabricated and used to measure temperatures in different fluids with a superior spatial resolution. An important task for the future will be the enhancement of the resolution. The achieved resolution of Pt/Au-thermocouple is relatively poor. By using semiconducting materials it should be possible to increase the thermoelectromotive force by one order of magnitude. One good candidate as material for the outer electrode will be tinoxide which can be easily doped and is also chemically inert. Another challenge will be the combination of a velocity sensor together with the temperature sensor in a very close vicinity. This gives the opportunity to measure the temperature and velocity at nearly the same position.

References

1. A. Tilgner, A. Belmonte, and A. Libchaber, Phys. Rev. E **47**, R2253 (1993).
2. Sheng-Qi Zhou and Ke-Qing Xia, Phys. Rev. Lett. **87**, 064501-1 (2001).
3. X.-L. Qiu and P. Tong, Phys. Rev. E **64**, 036304-1 (2001).
4. Sheng-Qi Zhou and Ke-Qing Xia, Phys. Rev. Lett. **89**, 184502-1 (2002).
5. X.-L. Qiu and P. Tong, Phys. Rev. E **66**, 026308-1 (2002).
6. X.-D. Shang, X.-L. Qiu, P. Tong, and K.-Q. Xia, Phys. Rev. Lett. **90**, 074501-1 (2003).
7. X.-L. Qiu, X.-D. Shang, P. Tong, and K.-Q. Xia, Phys. Fluids **16**, 412 (2004).
8. G. Fish, O. Bouevipch, S. Kokotov, K. Lieberman, D. Palanker, I. Turovets, and Aaron Lewis, Rev. Sci. Instrum. **66**, 3300 (1995).

Temperature and Velocity Measurements in a Large–Scale Rayleigh-Bénard Experiment

C. Resagk[1], R. du Puits[1], F.H. Busse[2], A. Thess[1], and A. Tilgner[3]

[1] Dept. of Mech. Eng., Ilmenau Univ. of Technology, 98684 Ilmenau, Germany
[2] Institute of Physics, Univ. of Bayreuth, 95440 Bayreuth, Germany
[3] Institute of Geophysics, Univ. of Goettingen, 37075 Goettingen, Germany

1 Introduction

The objective of our paper is to investigate the temperature and the velocity field in a large–scale Rayleigh–Bénard (RB) experiment at high Rayleigh number (Ra). The RB experiment, a closed box with a heated bottom plate and cooled top plate, offers good conditions for a systematic study of thermal convection. Measurements with high spatial and temporal resolution, especially in the boundary layers, can help to verify predictions of various theoretical models concerning the mechanism of the heat transport by highly turbulent thermal convection.

In previous RB experiments at high Ra usually liquid helium or water was used with the consequences, that the spatial dimensions and the thickness of the boundary layers are relatively small. The new investigations at the large–scale RB experiment allow precise measurements both at high Ra and at high aspect ratio Γ.

Two different experimental tasks are to be solved: first the measurement of velocity and temperature profiles in the boundary layer near the cooling plate and the determination of mean value data, and second the recording of long–term time series of velocity and temperature with the computation of power density spectra and autocorrelation function.

2 Experiment

The results were obtained in the presence of passive side walls with a thickness of 160mm, that do not allow a precise control of the thermal boundary conditions. Therefore all results have preliminary character. In the case of velocity and temperature profile measurements near the center line of the cylindrical convection cell these influence can largely be neglected.

In the present RB experiment with a variable height between $h = 0.1m$ and $6.30m$ and a diameter of $d = 7.15m$ we are able to study the velocity and the temperature field in great detail. We investigate profiles as well as the statistical properties of both fields in the boundary layer of the air–filled convection cell.

The velocity measurements were performed with a 1d Laser–Doppler–Anemometer (LDA) through a small glass window located in the center of the cooling plate. Two laser beams emitted from the LDA probe (Polytec) pass the window and intersect in the measurement volume. Tracer particles, inserted into the convection cell by a fog generator, scatter the laser light from the measurement volume back to the receiver of the LDA probe. The frequency of intensity modulated back-scattered light is proportional to the velocity component perpendicular to the optical axis in the plane of incidence.

Because of the small size of the measurement volume ($l_{mvz} = 0.250mm$, $d_{mvx} = 0.050mm$) in comparison to the thickness of the viscous boundary layer, the spatial resolution of the velocity profile is higher than in previous RB experiments. The LDA burst signal rate depends on the concentration of white fog particles and varies between 1Hz and 20Hz.

In order to measure the velocity profile the LDA probe is moved by a PC controlled traverse system with a positioning accuracy of 0.01mm in vertical (z) direction through the boundary layer up to a depth of 425mm. The distance between the 70 measuring points amounts to 0.1mm close to the wall and to 25mm in the bulk. The measuring time for each position averages 35min, for the complete profile 2 days. The 1d LDA probe measures the velocity in the dominant direction first, which we define as mean flow. After each profile recording the LDA probe is turned at 90° to obtain the y-component, which we call cross flow.

For the temperature measurement a glass encapsulated NTC thermistor (Thermometrics) with a diameter of $d_{th} = 0.140mm$ and a response time of 120ms is used. The temperature probe is moved in vertical direction at the same position and with equal measuring coordinates as the LDA measurements. Together with a computer controlled measurement system based on a HP3458 multimeter (8.5 digit, 100000 readings/s) temperature fluctuations down to the smallest scales are detectable.

The measurement system allows a recording of temperature time series with a sampling rate up to 333 Hz for each point. A recording time of 35min leads to a total number of temperature data of 700000 per measurement point.

The complete set of experimental parameters for the velocity and temperature measurements is listed in Tab.1.

3 Results and Discussion

3.1 Velocity Field Structure

One of the most important but until now unsolved problems is the characterization of the large-scale circulation (wind) in the bulk region of the convection

Table 1. Experimental parameters of velocity and temperature measurements.

Ra	10^{10}	10^{11}	10^{12}
Temperature of heating plate	26.0°C	37.2°C	71.2°C
Temperature of cooling plate	19.9°C	20.1°C	22.0°C
Height of cell	2.62m	4.00m	6.30m
Aspect ratio Γ	2.73	1.79	1.13

cell. In view of the mean flow hypothesis by Tilgner et al. [1] and Qui and Tong [2] the investigations should answer the question whether the predicted coherent single-roll structure still exist at an aspect ratio $\Gamma = 2$ or goes over into a multi-roll flow at higher aspect ratio $\Gamma > 2$.

The measured long-term velocity data with a duration of 48h per profile show a very different behavior dependent on Ra and Γ. At $Ra = 10^{12}$ and $\Gamma = 1.13$ we can clearly recognize a stable velocity profile in the viscous boundary layer. The constant direction of the mean flow over a period of 48h indicates the existence of a stable single roll.

By increasing Γ to 1.79 and decreasing Ra to 10^{11} the uniformity of the mean flow direction is broken and no well defined velocity profile can be measured. This tendency culminates in a total breakdown of the coherent structure at $Ra = 10^{10}$ and $\Gamma = 2.73$. In this case the steady mean flow velocity component is completely substituted by large fluctuations in all directions.

3.2 Temperature Field Structure

The measured temperature profiles are overall smoother than the velocity profiles. There is little dependence of the profiles on Ra and Γ. In all cases we get typical boundary layer profiles with a thermal boundary layer thickness of 1.6mm, 1.7mm and 1.8mm at $Ra = 10^{12}$, $Ra = 10^{11}$ and $Ra = 10^{10}$ respectively. These values are determined by the distance at which the extrapolation of the linear portion of the mean profile equals the central mean temperature.

The comparison of temperature fluctuations and velocity profiles permits an estimation of the viscous boundary layer thickness. Belmonte et al. [3] described a method based on the determination of the cutoff frequency in the temperature spectra. The highest cutoff frequency was found in a spectrum at a position where the standard deviation of velocity has a maximum.

The investigation at the convection cell at $Ra = 10^{12}$ shows a viscous boundary layer thickness of 20mm determined by cutoff frequency, of 22mm given by the maximum of the velocity standard deviation and 25mm measured in the mean velocity profile. These results confirm previous measurements of the Libchaber group [1].

3.3 Coherent Oscillations

Coherent structures in Rayleigh-Bénard convection were objectives of several experiments. Castaing et al [4] and Villermaux [5] describe them as a delayed coupling of boundary layer instabilities caused by hot and cold plumes. Our analysis of $v_x(t)$ and $v_y(t)$ should test this model and investigate the dynamic behavior of velocity and temperature oscillations.

The autocorrelation functions of mean and cross flow velocity at $Ra = 10^{12}$ show characteristic time scales of 20s and 40s. The dominant time scale of the temperature signal corresponds also 40s and agrees with the turnover time of the convection cell. Additionally to the short-term fluctuations one can find a long-term oscillation with a period of 500s.

The physical interpretation of the measured coherent oscillations requires further experimental investigations.

4 Conclusions

Our large-scale RB experiment permits high-resolution local measurements of velocity and temperature at $Ra >> 1$. The velocity data at $Ra = 10^{12}$ and $\Gamma = 1.13$ indicate the existence of a turbulent boundary layer.

The detection of a single-roll mean flow structure at $\Gamma = 1.13$ is in a good agreement with the theoretical predictions. For higher aspect ratio the single mean flow roll appears to breakdown completely.

Coherent structures are detectable in both velocity and temperature time series at $Ra = 10^{12}$ and $\Gamma = 1.13$.

At the present time our RB experiment is being upgraded by the installation of a counterheating system. Thus future results will be obtained with more precise boundary conditions.

Acknowledgments

We gratefully acknowledge the support of the Deutsche Forschungsgemeinschaft (grant number TH 497/16-1) and of the Thueringer Ministerium fuer Wissenschaft, Forschung und Kunst.

References

1. Tilgner A, Belmonte A, Libchaber A (1993) Phys Rev E 47: R2253–R2256
2. Qiu X L, Tong P (2001) Phys Rev E 64: 36304-1-36304-12
3. Belmonte A, Tilgner A, Libchaber A (1994) Phys Rev E 50: 269-279
4. Castaing B, Gunaratne G, Heslot F, Kadanoff L, Libchaber A, Thomae S, Wu X Z, Zaleski S, Zanetti G (1989) J Fluid Mech 204: 1-30
5. Villermaux E (1995) Phys Rev Lett 75: 4618-4621

Statistics and Scaling of the Velocity Field in Turbulent Thermal Convection

Ke-Qing Xia, Sheng-Qi Zhou, and Chao Sun

Department of Physics, The Chinese University of Hong Kong, Shatin, Hong Kong, China. kxia@phy.cuhk.edu.hk

The statistics and scaling properties of the velocity field in turbulent Rayleigh-Bénard convection in water has been measured using both laser Doppler velocimetry (LDV) and particle image velocimetry (PIV) techniques. It is found that results from both techniques for the mean velocity and all the statistical quantities examined agree with each other. The measurements reveal that the pdfs for the velocity are non-Gaussian in the cell center but more close to Gaussian near the cell boundaries. In addition, the Reynolds shear stress is found to have different signs near the sidewall and near the plates of the cell, suggesting that different mechanisms are responsible for driving the mean flow at different locations of the cell. Moreover, our results confirm a prediction of a recent model by Grossmann & Lohse, in which flow geometries are classified according to the shape of the container.

1 Introduction

In 1987, Heslot, Castaing and Libchaber[1] carried out a high precision convection experiment in low temperature gaseous helium. Based on the properties of global heat transport and local temperature field, the Chicago group divided the Rayleigh-Bénard convection in the turbulent regime into the soft turbulence state (Ra $= 5\times10^5 \sim 4\times10^7$) and the hard turbulence state (Ra $\geq 4\times10^7$). The division of soft and hard turbulence is an important contribution to the Rayleigh-Bénard convection problem[2]. To understand the hard turbulence regime in Rayleigh-Bénard convection, many experimental, numerical, and theoretical efforts have been made in the last decade or so. For a recent review on this subject, please see Grossmann and Lohse[3], and Kadanoff[4]for a more pedagogical description, on the outstanding issues.

Despite many experimental studies of convective thermal turbulence since the work of the Chicago group, many fundamental quantities that are important in "ordinary" hydrodynamic turbulence systems have not been measured in the Rayleigh-Bénard system. Very recently, we have measured for the first time many quantities of the two-dimensional velocity field in a rectangular shaped convection box[5], using the PIV technique. These include the mean velocity field, rms velocity field, skewness field and the Reynolds stress field. The main advantages of the PIV method are its ability to make instantaneous velocity measurement for the whole flow field and making such measurement equal-spaced in the time domain. But even for measuring time-averaged flow properties, PIV's ability of making simultaneous whole-field measurement greatly reduces the time required for determining the spatial structures of the velocity field. On the other hand, the large amount of information produced by the whole-field measurement also gives rise to the main disadvantage of the PIV, which is low data rate and short data acquisition time as compared to the LDV method. This is caused by the relatively low data transfer rate between the CCD camera and the computer and the limited memory space of the computer. Because of the very limited data rate of the PIV technique, it is difficult to obtain a large data record, which makes it impossible to examine the higher order statistical quantities. In this paper we report measurements of the statistical and scaling properties of the velocity field using both the LDV and PIV techniques, the results show that the two methods are complimentary to each other in the study of convective turbulence.

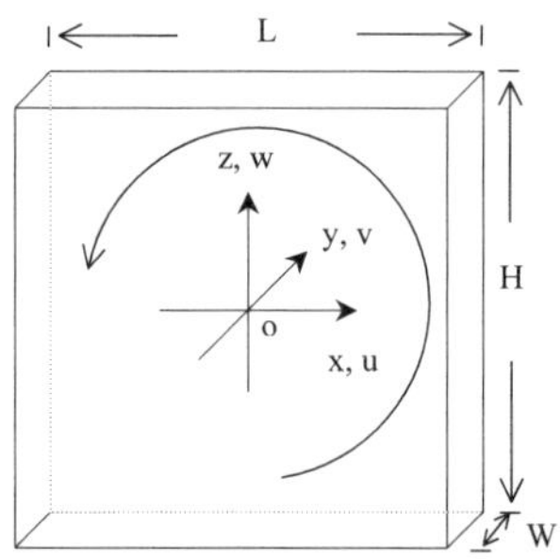

Fig. 1. Illustration of the convection cell and the coordinates of the experiment.

2 Experiment

Figure 1 shows a simple illustration of the convection cells used in the present work, which have been described in detail elsewhere[5]. Two cells are used, the dimensions, in unit of cm, of the large cell are $L = 81$, $H = 81$ and $W = 20$ and those of the small one are $L = 25$, $H = 24$ and $W = 7$. Also shown in the figure is the experimental coordinate system, which is so defined that its origin coincides with the cell center, its x-axis points to the right, the z-axis points upward and the y-axis points inward. Given the geometry of the cell, the large-scale mean flow is expected to be confined mainly within the (x, z) plane, where the relevant aspect ratio for the flow is unity. The Rayleigh number is defined as $\mathrm{Ra} = \alpha g H^3 \Delta T/\nu\kappa$, with g being the gravitational acceleration, and α, ν, and κ, respectively the volume expansion coefficient, the kinematic viscosity, and the thermal diffusivity of the convecting fluid. The Prandtl number is defined as $\mathrm{Pr}=\nu/\kappa$. In the results presented below, the LDV data were taken at $\mathrm{Ra} = 5.4\times10^{11}$ in the large cell and the PIV measurement was made in both cells with Ra spanning from 9×10^8 to 9×10^{11}, and the Prandtl number $\mathrm{Pr} \simeq 4.3$ in all cases.

A two-component commercial LDV system (Dantec Ltd.) was used in the local velocity measurement. The size of the LDV measuring volume was approximately 0.087 mm in diameter and 1.11 mm in length in water and the fluid was seeded with neutrally buoyant 2.2-μm diameter latex spheres as scatterers. The data rate of the experiment was between 20−100 Hz, depending on the location of measurement. As the typical cutoff frequency in our system is only a few Hz, this data rate ensured that the smallest scales were detected reliably. At every measuring point 1,048,576 statistically independent samples, or, alternatively, at least 5.2 hour-long time series, were acquired. The PIV system and the measurement details have been described previously in [5]. Briefly, it consists of a dual Nd:YAG laser, a lightsheet optics, a CCD camera and a synchronizer. The seeding particles used in the experiment are 50 μm diameter polyamid spheres. A 2D velocity map is obtained by cross-correlating two consecutive images separated by a time interval ranging from 20 to 100 ms that is selected optimally according to the flow speed. The mean flow field is an average of 15,000 vector maps acquired at a sampling rate of 3 Hz.

3 Results and Discussion

3.1 Statistical Properties

As already mentioned, many quantities in the convection cell have been measured by the PIV technique and have been reported previously[5]. As a cross-check between the results from the PIV measurement and the LDV measurement, we show in Fig. 2 the probability density function (pdf) of the vertical velocity w and the horizontal velocity u at various positions along the x-axis (the x-scan), obtained respectively by the two methods: the left panel are those from LDV and the right from PIV. It is seen from the figure that the pdf's obtained by the two methods show qualitatively the same features. In fact all the statistical quantities, such as the rms, skewness and flatness values obtained from the two techniques are qualitatively all the same for comparable values of Ra. Thus, in the rest of Sec. 3.1, only LDV results will be presented. From Fig. 2 we also see that the LDV result produces better statistics than the PIV due to a much-longer data record. Fig. 2 reveals that the velocity

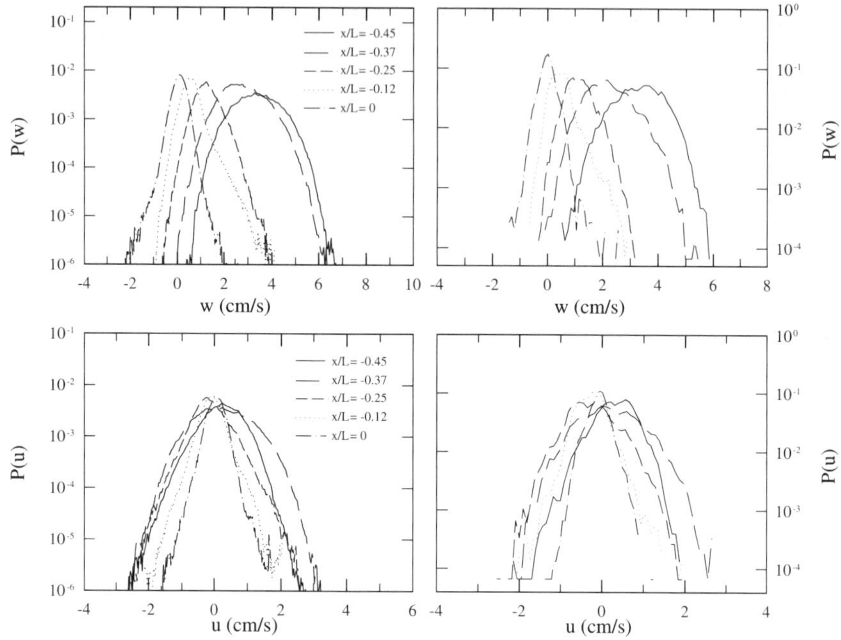

Fig. 2. Probability density functions of w (upper panel) and u (lower panel) at several locations along the x-axis; Left: LDV data (Ra $= 5.4\times10^{11}$). Right: PIV data (Ra $= 3.5\times10^{10}$).

pdfs for both the vertical and horizontal velocity components at various positions change their shape from near Gaussian to nearly exponential as the location moves from near sidewall to the cell center, similar features are also seen for the z-scan (i.e. from the bottom plate to the cell center along z-axis). Velocity pdfs have been measured previously by Qiu & Tong[8] in a cylindrical cell, in which they observed that the pdf taken at the cell center is more close to Gaussian than at other locations. Daya and Ecke have reported previously that the scaling properties of both the fluctuating temperature and velocity fields can depend on the shape of the convection box[9]. It is thus not surprising that the velocity statistics in the rectangular and cylindrical cells are different, but it is surprising that in the rectangular case the pdf is non-Gaussian in the center but Gaussian near the boundary.

We now examine in Fig. 3 the profiles of the skewness, $S_u = \langle(u-U)^3\rangle/u^3_{rms}$ and $S_w = \langle(w-W)^3\rangle/w^3_{rms}$, and flatness $F_u = \langle(u-U)^4\rangle/u^4_{rms}$ and $F_w = \langle(w-W)^4\rangle/w^4_{rms}$, of the horizontal and vertical velocities respectively, here $U = \langle u\rangle$ and $W = \langle w\rangle$. Note that the value of the flatness should be 3 for a Gaussian distribution. Fig. 3 thus shows that the velocity distribution is close to Gaussian only in a small region near the top and bottom plates and near the sidewalls, as already seen from Fig. 2. We have previously found that an inner core-outer shell structure exists in the central of the convection

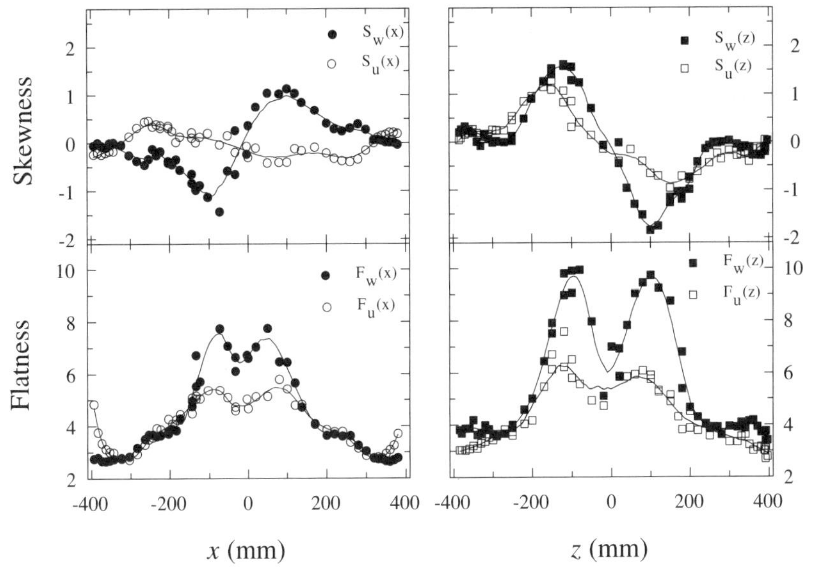

Fig. 3. Skewness and flatness of the velocity components u and w measured in the x-scan (left panel) and z-scan (right panel).

box[5]. Here we note that the peak positions of the skewness and flatness correspond approximately to the boundary between the inner-core and outer shell. This feature has already been discovered by the PIV measurement[5], now the "profile-cut" from LDV measurement confirms it and makes it more explicit.

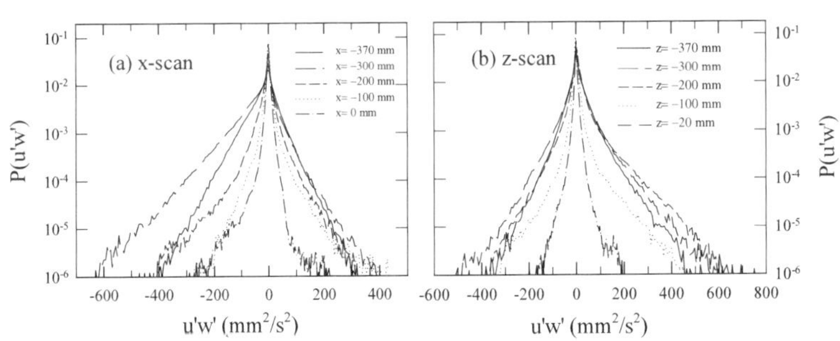

Fig. 4. Probability density function of $u'w'$ measured at several locations along (a) x-scan and (b) z-scan.

With the simultaneous measurement of the two velocity components, we can obtain the Reynolds stress, which is responsible for transporting energy from the mean flow to the fluctuating turbulent flow and thus plays a significant role in turbulent production. Therefor, a study of its characteristics

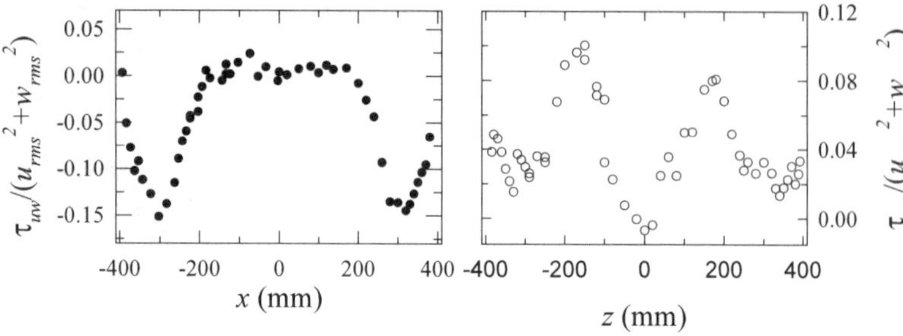

Fig. 5. Profiles of the normalized Reynolds shear stress along the x-scan (left panel) and z-scan (right panel).

is of great importance in understanding the mechanism of turbulent energy transfer. Its role has been studied extensively in the channel, jet, shear flows, and in boundary layers. However, few publications are available on the role of the Reynolds stress in thermal turbulence. Without the third velocity component v, we can discuss only one shear stress component $\tau_{uw} = \langle u'w' \rangle$ here, which is the dominant component of the Reynolds shear stress tensor in the present case. Here $u'(t) = u - U$ and $w'(t) = w - W$ are the fluctuating parts of u and w. In [5], we have measured the time-averaged two-dimensional Reynolds stress field and found that regions of large τ_{uw} correspond to zones where the hot and cold plumes start to accelerate along the sidewalls toward the opposite plates. With the high sampling frequency and long data acquisition time afforded by the LDV measurement, we now examine the statistics of the "instantaneous Reynolds stress" tensor $u'(t)w'(t)$. Figure 4 plots the pdf of $u'w'$ at several positions along the horizontal and vertical axes. The figure shows that the Reynolds stress is more intermittent than the individual velocity components as seen in Fig. 2, it also shows that $u'w'$ near the perimeter of the cell has much larger magnitude of fluctuations than in the central region. Note also that $u'w'$ is skewed towards positive near the plates and towards negative near the sidewalls. This feature can also be seen more explicitly from the profiles of the normalized Reynolds stress $\tau_{uw}/(u_{rms}^2 + w_{rms}^2)$ shown in Fig. 5. When taking into account the sign of the mean shear (which can be obtained from the mean velocity profiles[5]), this implies that the turbulent energy production $P = -\langle u_i u_j \rangle \partial U_i / \partial x_j$ is negative in the regions near conducting plates, whereas it is positive near the sidewalls. The production dictates energy transfer between the mean flow and turbulence, a negative production means the mean flow obtains energy from turbulent fluctuations while a positive one implies that the mean flow supplies energy to the turbulent fluctuations, which is the typical situation for most flows. Therefore, near the sidewall turbulent fluctuations get their energy from the mean flow which in turn are driven by buoyancy through the accelerating plumes. But near the plates turbulent fluctuations provide energy to the mean flow which is how the horizontal part of the mean flow obtains its energy. It has been suggested long ago that the mean flow is driven by Reynolds stresses associated with tilting plumes[6], which implies a negative production P. Here we see that on the horizontal plates, velocity fluctuations do indeed provide energy to the mean flow, although P is overall positive in the cell, which means on the whole the mean flow is not driven by fluctuations, as already found by both the authors[5] and others[7].

3.2 Velocity Scaling in Confined Flow Geometries

It has been previously found that the viscous boundary layer at the thermally insulating sidewall[10] is different in terms of its scaling properties from that near the conducting plates[11, 12]. While it is expected that thermal boundary layers should be different at the plates and at the sidewall, the fact that the corresponding viscous boundary layers are also different came as a surprise, as both are created by the shear of the same large-scale flow that circulates around the periphery of the convection box. Based on volume flux conservation of the mean flow, Grossmann and Lohse (GL) have put forward a model to account for this discrepancy, in which the flow geometry is classified into two cases according to the shape of the container: (a) laterally restricted flow and (b) plate filling flow[13]. Experimentally, flow in our rectangular cell corresponds to the former, while cylindrical and cubical cells would correspond to the latter. A specific result of the GL model is that velocity near the sidewall and the plates is the same if the aspect ratio in the flow plane is unity. With the PIV-measured velocity field in the rectangular box, this can be checked directly. In Fig. 6 we plot the normalized velocity $U_w H/\kappa$ and $U_p H/\kappa$ against Ra, where U_w and U_p are the velocities at positions 2 cm from the left sidewall at mid-height and 2cm above the center of the bottom plate, respectively. The results show clearly that the velocity near the sidewall and near the plates are the same both in magnitude and in the scaling with Ra for flow in the (x, z) plane in which the *laterally restricted* flow geometry applies.

4 Summary and Conclusion

We have measured the statistics and scaling properties of the velocity field in high-Rayleigh number turbulent thermal convection in water, using rectangular-shaped containers. Both the LDV and the PIV techniques are used in the experiment. It is found that results from both techniques for the mean velocity and all the statistical quantities examined agree with each other. Specifically, we found that the pdfs for the velocity are non-Gaussian in the cell center but more close to Gaussian near the cell boundaries, which is in sharp contrast to the findings for convection in cylindrical cells. The velocity skewness and flatness results reveal that velocity fluctuations in the convection box are highly anisotropic and spatially inhomogeneous. Moreover, the position of the inner-core and outer-shell boundary is found to be a region with large values

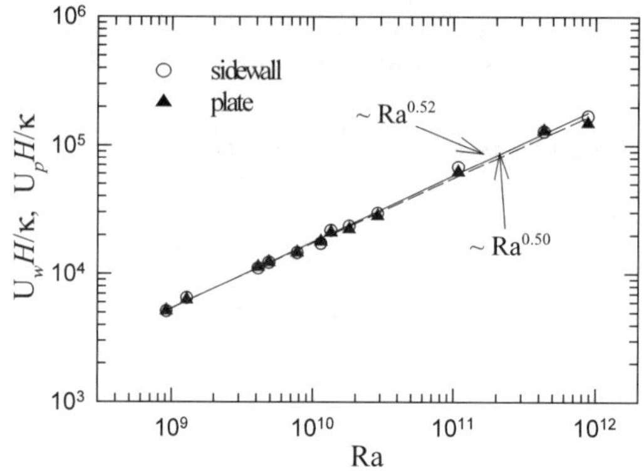

Fig. 6. Ra-dependence of the normalized velocity U_wH/κ and U_pH/κ from near-wall and near-plate positions, respectively.

of skewness and flatness. The results from Reynolds shear stress suggest that different locations of the cell have mechanisms for driving the mean flow. Finally, the velocity measurements directly confirm a recent prediction of the model by Grossmann & Lohse that classifies flow geometries according to the shape of the container.

We are pleased to acknowledge support of this work by the Hong Kong Research Grants Council under Project No. CUHK4242/01P and by CUHK Research Committee under Project No. 2060229.

References

1. Heslot F., Castaing B., Libchaber A. (1987) Phys. Rew. A 36 (12):5870-5873.
2. Castaing B., Gunaratne G., Heslot F., Kadnoff L.P., Libchaber A., Thomae S., Wu X.Z., Zaleski S., Zanetti G. (1989) J. Fluid Mech. 204:1-30.
3. Grossmann S, Lohse D (2000) J. Fluid Mech. 407:27-56.
4. Kadanoff L.P. (2001) Phys. Today 54 (8):34-39.
5. Xia K.Q., Sun C., Zhou S.Q. (2003) Phys. Rev. E 68:066303(1–18).
6. Krishnamurti R., Howard L. N. (1981), Proc. Natl. Acad. Sci. U.S.A. 78:1981–1985.
7. Burr U., Kinzelbach W., Tsinober A. (2003) Phys. Fluids 15:2313–2320.
8. Qiu X.-L., Tong P. (2001) Phys. Rev. E 64:036304(1–13).
9. Daya Z.A., Ecke R.E. (2001) Phys. Rev. Lett. 87: 184501(1-4).
10. Qiu X.-L.,Xia K.-Q. (1998) Phys. Rev. E 58:486–491.
11. Xin Y.B., Xia K.-Q., Tong P. (1996) Phys. Rev. Lett. 77:1266–1269.
12. Lam S., Shang X.-D., Zhou S.-Q., Xia K.-Q. (2002) Phys. Rev. E 65:066306(1–8).
13. Grossmann S., Lohse D. (2003) J. Fluid Mech. 486:105–114.

Part V

Simulation (DNS and LES)

Nonlinear Stochastic Estimation: A Tool for Deriving Appropriate Wall Models for LES

M. Abel[1], D. Stojković[2], and M. Breuer[2]

[1] Institute of Physics, University of Potsdam, Am Neuen Palais 19, 14415 Potsdam, markus@stat.physik.uni-potsdam.de
[2] Institute of Fluid Mechanics, University of Erlangen, Cauerstr. 4, 91056 Erlangen, breuer@lstm.uni-erlangen.de

Wall Models for LES

In LES, a key technology for the application to complex high–Re flows is an appropriate wall modeling strategy. The credibility and costs of LES results strongly depends on these wall models. In this contribution, we show how to apply an a priori nonparametric estimation procedure to find generalized wall models. In the approach of Schumann [1], the wall shear stresses $\tau_{w,ij}$ and the velocity components in the first off-the-wall cell have been correlated directly. The resulting system of model equations reads:

$$\tau_{w,xy} = \langle \tau_w \rangle / \langle \bar{u}(Y) \rangle \, \bar{u} \ , \quad \tau_{w,yy} = 0 \ , \quad \tau_{w,zy} = Re^{-1} \, \bar{w}/Y \ , \tag{1}$$

where x, y, z, t denote the streamwise, wall-normal, and spanwise coordinate and time; $\tau_{w,ij}$ depends on (x, z, t) and the filtered streamwise and spanwise velocity components, $\bar{u}$ and $\bar{w}$ depend on (x, Y, z, t) with Y the distance to the wall. Brackets, $\langle \cdot \rangle$, denote spatio-temporal averaging. This model contains no adjustable parameters, but a slope which varies with $\langle \tau_w \rangle / \langle \bar{u}(Y) \rangle$. In the following non-dimensional quantities are used:
$\tau_w \mapsto \tau_w / \langle \tau_w \rangle$, $u \mapsto \bar{u}/\langle \bar{u}(Y) \rangle$, $v \mapsto \bar{v}/\langle \bar{u}(Y) \rangle$, $\partial_x p \mapsto \partial_x \bar{p} / (\rho \langle \bar{u}(Y) \rangle)^2 / \delta_{Ch})$.

Many modifications and extensions of Schumann's model have been proposed. We use an algebraic model, which includes the models proposed in [2] ("Piomelli's model" and "ejection model"). Schumann's model relates τ_w and u. Piomelli's model takes into account inclined structures near the wall by a streamwise shift, $x \mapsto x + \delta$, such that $\bar{u}$, $\bar{v}$, $\bar{w}$ depend on $(x + \delta, Y, z, t)$. The ejection model includes a dependency on the wall-normal velocity. We additionally suggest a dependence of $\tau_{w,xy}$ on $\partial_x p$ and $\tau_{w,zy}$ on $\partial_z p$. A non-parametric, additive model then reads:

$$\tau_{w,xy} = f_{u,1}\left(u(\mathbf{r}, t)\right) \; + \; f_{v,1}\left(v(\mathbf{r}, t)\right) \; + \; f_{p,1}\left(\partial_x p(\mathbf{r}, t)\right) \; + \; \epsilon_1 \ , \tag{2}$$

$$\tau_{w,zy} = f_{u,3}\left(u(\mathbf{r}, t)\right) \; + \; f_{v,3}\left(v(\mathbf{r}, t)\right) \; + \; f_{p,3}\left(\partial_z p(\mathbf{r}, t)\right) \; + \; \epsilon_3 \ , \tag{3}$$

and additionally $\tau_{w,yy} = 0$; here u, w depend on $(\mathbf{r}, t) = (x + \delta, Y, z, t)$ and $\epsilon_{1/3}$ denotes the model error with mean value zero. Below, we show that our analysis allows to measure the importance of modeling terms $f_{\cdot,i}$. The addition of terms is understood as a tentative input to the analysis.

Nonparametric Stochastic Estimation

In stochastic estimation a least-square minimization procedure is applied to data taken from DNS or measurements [3]:
$E\left[\epsilon^2\right] = E\left(\left[\tau_w(\mathbf{r},t) - \hat{f}_u\left(u(\mathbf{r},t)\right)\right]^2\right) = \min$. Here, $\hat{f}$ is the optimal estimate for f_u. If $f_u(u)$ is linear or parametrized ,e.g. polynomial, one can use linear regression tools. Consequently, the procedure has been called **L**inear **S**tochastic **E**stimation, LSE [3]. If the function f_u is a priori unknown, one can assume a nonparametric, possibly nonlinear function. Now the minimization is achieved by nonparametric regression [4]: The conditional expectation value operator $E(\tau_w|u)$ [3] is applied. Then we obtain for $\hat{f}_u$ (the hat is omitted below):

$$\hat{f}_u(u) = E(\tau_w|u) = \int P(\tau'_w|u)\tau'_w \, d\tau'_w \ , \tag{4}$$

where $P(\tau'_w|u)$ denotes the conditional probability. Equivalently, one can maximize the correlation, C, used as criterion below. If additive models of the type (2) or (3) are considered, the estimation procedure is slightly more complicated and of iterative nature, since the problem is intrinsically nonlinear [4, 5]. To obtain a criterion for the importance of a single term, say $f_{u,1}$ in (2), one calculates $\tau_{w,xy} - f_{v,1} - f_{p,1}$ and correlates with $f_{u,1}$:

$$C_1\left[f_{u,1}, \ (\tau_{w,xy} - f_{v,1} - f_{p,1})\right] \ . \tag{5}$$

Data

We use instantaneous DNS and wall-resolved LES data from the second-order, finite-volume code $\mathcal{LESOCC}$ [6]. For comparison with other results, we consider the plane channel flow. To investigate Reynolds number effects, data are generated for $Re = 2800$, $Re = 4000$ (DNS) and $Re = 10,935$, $Re = 22,000$ (wall-resolved LES). For the analysis a sub-domain in x, y, z has been taken over 25 characteristic times, L_y/U_b. Details about the simulation, the channel geometry, the data domain are found in [5] .

Results

We present the main NLSE results concerning different Re, NLSE/LSE correlations and optimal functions. The analysis concerns (2) and (3). We will focus on the first and briefly comment on the latter. Since Schumann's, Piomelli's and the ejection model are subsets of (2), results for the respective

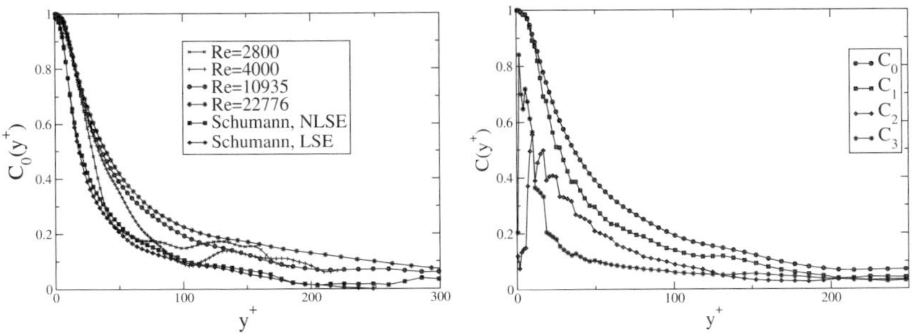

Fig. 1. a) Correlations C_0 for the gradient model and Schumann's model, NLSE and LSE ($Re = 10,935$), b) Correlations C_0–C_3 for $Re = 10,935$.

nonparametric generalizations can be read from the gradient model results. For comparison, we plot correlations C against $y^+ = y \cdot Re \cdot \langle u_\tau \rangle$ for all Reynolds numbers. The gradient model involves 4 terms, so we calculate 4 correlations $C_0 = C(\tau_{w,xy}, f_{u,1} + f_{v,1} + f_{p,1})$, $C_1 = C(f_{u,1}, \tau_{w,xy} - f_{v,1} - f_{p,1})$, $C_2 = C(f_{v,1}, \tau_{w,xy} - f_{u,1} - f_{p,1})$, and $C_3 = C(fp, 1, \tau_{w,xy} - f_{u,1} - f_{v,1} - f_{p,1})$ (see (5)). The details about C_0–C_3 and the optimal functions are shown for $Re = 10,935$. The other data sets yield similar results.

Correlations and Importance

From Fig. 1a), one reads that the correlations C_0 increase with Reynolds number. This holds as well from $Re = 2800$ to 4000, so we conclude that it is not a result of DNS and LES differences. For Schumann's model, LSE yields slightly smaller correlation than NLSE; the bigger impact comes from the addition of a delay. Figure 1b) shows that the major contribution comes from $f_{u,1}$ (C_1, Piomelli's model). The function $f_{v,1}$ is unimportant close to the wall, gains importance until a maximum at $y^+ \simeq 18$ to become unimportant again for larger distances (C_2, ejection model). Close to the wall $f_{p,1}$ is highly correlated (C_3). This hints to the inclusion of similar terms in models for separating flows (if the range of correlation extends to larger y^+).

Optimal Functions

A plot of the optimal functions for the gradient model provides further insight. For $f_{u,1}$ (Fig. 2a), generalized Piomelli's model) we find for small y^+ a basically linear function. For larger distances curvature occurs, indicating nonlinear behavior; for large distances, correlation between the wall shear stress and the velocity field is lost and $f_{u,1} \simeq const.$ Inspecting Fig. 2b) (ejection model) one finds roughly $f_{v,1} = \alpha \cdot v$ for $v \lesssim 0$ and $f_{v,1} = const.$ for $v \gtrsim 0$, corresponding to sweeps. A linear regression is unable to find these details. With increasing

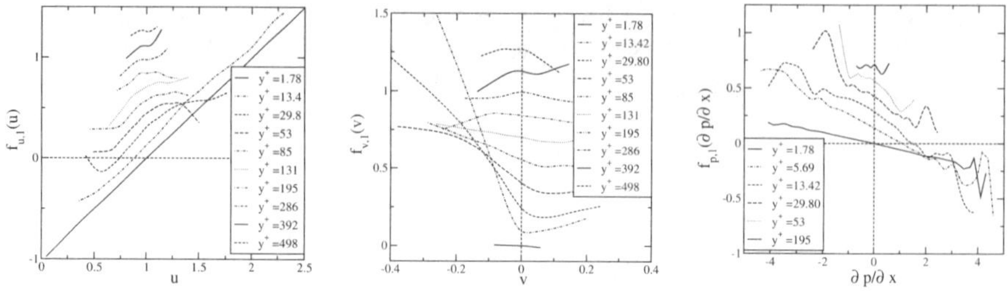

Fig. 2. Optimal functions for the gradient model: (a) $f_{u,1}$ (b) $f_{v,1}$, and (c) $f_{p,1}$ (Re=10,935).

y^+, $f_{v,1}(v) \to const.$ due to decorrelation. The function of the streamwise pressure gradient (Fig. 2c) is approximately linear with negative slope for very small y^+ and goes over a concave shape to a constant for $y^+ \gtrsim 20$. In brief, the results for the NLSE of (3) ($\tau_{w,zy}$) are: the $f_{u,3}$ term and the $f_{p,3}$ are important, the latter term even dominates for $y^+ > 20$. This confirms the intuition that the pressure gradient is important for the dynamics.

Conclusions

A priori analysis with NLSE yields clearly higher correlations for nonlinear and nonparametric modeling. For the implementation of models, one is, however, faced with the problem to find the correct asymptotics of generalized functions. The first results of implementations let us hope that for more detailed investigations, including correct filtering and possibly larger data sets, better results are obtained. We think, that the true strength of the method lies in its nonlinear character which can be used for models in separated flows.

References

1. U. Schumann. Subgrid–scale model for finite–difference simulations of turbulent flows in plane channels and annuli. *J. Comput. Physics*, 18:376–404, 1975.
2. U. Piomelli, J.H. Ferziger, P. Moin, and J. Kim. New approximate boundary conditions for large eddy simulations of wall–bounded flows. *Phys. Fluids A*, 1:1061–1068, 1989.
3. R. J. Adrian, B. G. Jones, M. K. Chung, Y. Hassan, C. K. Nithianandan, and A. Tung. Approximation of turbulent conditional averages by stochastic estimation. *Phys. Fluids A*, 1:992, 1989.
4. T.J. Hastie and R.J. Tibshirani. *Generalized Additive Models.* Chapman and Hall, London, 1990.
5. M. Abel, D. Stojković, and M. Breuer. Nonlinear stochastic estimation of wall functions for LES. submitted to Phys. Fluids.
6. M. Breuer. *Direkte Numerische Simulation und Large–Eddy Simulation turbulenter Strömungen auf Hochleistungsrechnern.* Habilitationsschrift, Universität Erlangen–Nürnberg. Shaker Verlag, Aachen, 2002.

Generation of Mean Flows in Turbulent Convection

T. Hartlep and A. Tilgner

Institute of Geophysics, University of Göttingen, Herzberger Landstr. 180, 37075 Göttingen, Germany

Summary. The generation of mean flows in horizontally periodic Rayleigh-Bénard convection is studied using direct numerical simulations (DNS) with Rayleigh numbers up to 10^7. We present the spatial and temporal characteristics of this flow component and its dependence on the parameters of the problem.

1 Introduction

Thermal convection is a very common phenomenon in fluid mechanics and plays an important role in geophysical systems like the atmosphere and the oceans, as well as in engineering applications. The Rayleigh-Bénard problem in particular deals with thermal convection in a horizontal fluid layer that is heated from below and cooled from above. It has long been one of the preferred systems for the study of turbulence both experimentally, numerically and theoretically. This is last but not least due to its rather localized flow structures. On the largest scales, these flows are usually characterized by roll-like or cellular patterns, which seem to persist up to the highest Rayleigh numbers, as atmospheric observations suggest. It is known from laboratory experiments [3] and numerical simulations [5] that the horizontal size of these structures is of the order 1–10 times the layer depth and increases with Ra. The question this paper addresses is whether thermal convection can in addition to these flow patterns produce a large scale mean flow, as the experiments by Krishnamurti and Howard [7] suggest.

2 Mathematical Formulation of the Problem

We consider the usual Rayleigh-Bénard problem: a fluid layer of depth d between two rigid horizontal plates of fixed temperatures. A positive temperature difference ΔT is maintained between the lower and the upper plate. Acceleration of gravity is acting in negative z direction. Periodic boundary

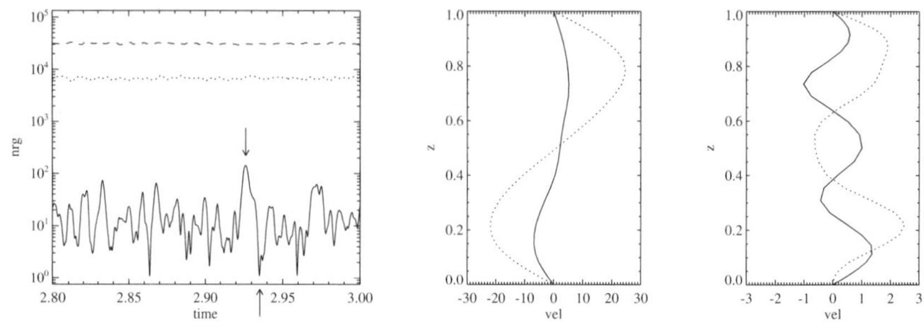

Fig. 1. Temporal evolution of mean flow (solid), toroidal (dotted) and poloidal (dashed) contributions to the kinetic energy in the case $Ra = 10^6$, $Pr = 7$, $\Gamma = 10$ (left figure). The two diagrams on the right show snapshots of mean flow profiles of the same simulation at times $t = 2.92629$ and $t = 2.93526$, where U_x and U_y are denoted by solid and dotted lines, respectively.

conditions are used in x and y directions with periodicity lengths l_x and l_y, defining the aspect ratio as $\Gamma = l_x/d = l_y/d$. We use the Boussinesq approximation in the governing equations. The relevant fluid properties are the kinematic viscosity ν, the thermal diffusivity κ, and the coefficient of thermal expansion α. Using d, d^2/κ, ΔT and $\rho\kappa^2/d^2$ as units of length, time, temperature, and pressure, one obtains the equations for the non-dimensional velocity $\mathbf{v}(x, y, z, t)$ and temperature $T(x, y, z, t)$:

$$\partial_t \mathbf{v} + (\mathbf{v} \cdot \nabla)\mathbf{v} = -\nabla p + Pr\ \nabla^2 \mathbf{v} + Ra\ Pr\ T\hat{\mathbf{z}}, \tag{1}$$

$$\partial_t T + (\mathbf{v} \cdot \nabla)T = \nabla^2 T, \tag{2}$$

$$\nabla \cdot \mathbf{v} = 0, \tag{3}$$

which contain two dimensionless control parameters: the Rayleigh number $Ra = g\alpha d^3 \Delta T/(\kappa\nu)$ and the Prandtl number $Pr = \nu/\kappa$. Since incompressibility is assumed, the velocity field can be uniquely represented by a poloidal scalar $\phi(x, y, z, t)$, a toroidal scalar $\psi(x, y, z, t)$ and a mean flow $\mathbf{U}(z, t)$ [9]:

$$\mathbf{v} = \nabla \times \nabla \times \phi\hat{\mathbf{z}} + \nabla \times \psi\hat{\mathbf{z}} + \mathbf{U}. \tag{4}$$

ϕ and ψ are bounded functions with vanishing averages over horizontal planes. The z component of $\mathbf{U}$ vanishes. The resulting equations are numerically solved by a spectral method [8, 6, 5, 4]. A fairly detailed description of the numerical method can be found in [4]. Initial conditions for all simulations was a fluid at rest with some random temperature disturbances.

3 Generation of Mean Flows

The notion "mean flow" in the context of this paper describes a horizontally averaged flow represented by the velocity component $\mathbf{U}$ of equation (4). This

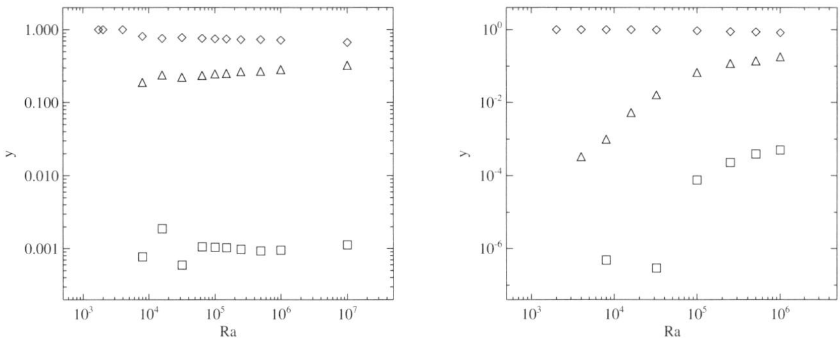

Fig. 2. Rayleigh number dependence of poloidal (diamonds), toroidal (triangles) and mean flow energy (squares) for $\Gamma = 10$ and Prandtl numbers $Pr = 0.7$ (left) and $Pr = 7$ (right). All energies are shown in fractions of the total kinetic energy.

of course is not to be confused with a time averaged flow, for which the term mean flow is sometimes used as well. Indeed, $\mathbf{U}$ is highly time dependent except for the most simple laminar cases. Several instabilities of laminar convection are known which generate mean flows. Examples include travelling wave convection at Prandtl numbers of order unity with a symmeric mean flow along the primary rolls [1], and the wavy oscillatory instability of bimodal convection which produces an antisymmetric mean flow [2]. Here, symmetric and antisymmetric refer to the symmetries with respect to the midplane of the fluid layer ($z = 0.5$).

As a typical example, figure 1 shows a section of the time evolution of the different contributions to the kinetic energy E_{kin} in the case of turbulent convection at $Ra = 10^6$, $Pr = 7$, $\Gamma = 10$. The time interval shown corresponds to approximately 50 convective time scales $\tau = (2E_{kin})^{-1/2}$. Note the rather small amplitude and high variability of the mean flow energy. Figure 1 also presents some vertical profiles of the mean flow $\mathbf{U}$ at two different times marked by the arrows in the left diagram. At a time when the mean flow contribution is relatively large one observes a mainly antisymmetric mean flow (middle diagram of figure 1). In fact, in all investigated $Pr = 7$ cases the antisymmetric part dominiated the mean flow. The same is true for $Pr = 0.7$ except for small Rayleigh numbers below $Ra = 3.2 \times 10^4$.

The Rayleigh number dependence of the kinetic energy distribution on the various velocity components is plotted in figure 2. Note a striking difference in behavior between $Pr = 0.7$ and $Pr = 7$. The mean flow energy E_{mf} at the lower Prandtl number is roughly 0.1 percent of the total kinetic energy, independent of Rayleigh number. On the other hand, the $Pr = 7$ cases show a strong increase of the relative mean flow energy with Ra, although it seems that it might saturate at higher Rayleigh numbers. Further simulations are under way to give a definitive answer.

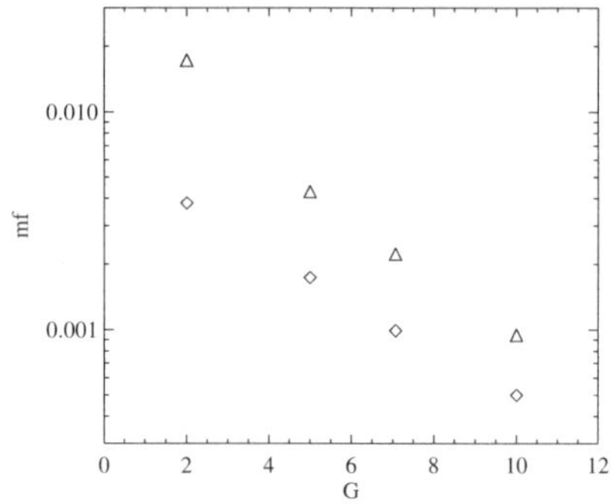

Fig. 3. Aspect ratio dependence of the mean flow energy E_{mf} for Rayleigh number $Ra = 10^6$ and Prandtl numbers $Pr = 0.7$ (triangles) and $Pr = 7$ (diamonds), again plotted in fractions of the total kinetic energy E_{kin}.

In the last figure we present some results on the aspect ratio dependence of the mean flow. A reduction of Γ strongly increases the kinetic energy of the mean flow. Still, even at the smallest aspect ratio E_{mf} accounts for less than 2 percent of the total energy (Pr = 0.7 case). The dominant flow structure is still the convection roll and not the mean flow.

References

1. Clever RM, Busse FH (1989) J Fluid Mech 201:507–523
2. Clever RM, Busse FH (1994) J Fluid Mech 271:103–118
3. Fitzjarrald DE (1976) J Fluid Mech 73:693–719
4. Hartlep T, Tilgner A (2003) In: High Performance Computing in Science and Engineering '03. Springer, Berlin Heidelberg New York
5. Hartlep T, Tilgner A, Busse FH (2003) Phys Rev Lett 91:064501
6. Kerr RM (1996) J Fluid Mech 310:139–179
7. Krishnamurti R, Howard, LN (1981) Proc Natl Acad Sci USA 78:1981–1985
8. Moser RD, Moin P, Leonard A (1983) J Comp Phys 54:524–544
9. Schmitt BJ, von Wahl W (1992) Diff Int Eqs 5:1275–1306

Control of a Turbulent Separation Bubble by Periodic Excitation

Astrid H. Herbst[1] and Dan S. Henningson[2]

[1] KTH, Mechanics
10044 Stockholm, Sweden herbsta@mech.kth.se
[2] KTH, Mechanics
10044 Stockholm, Sweden henning@mech.kth.se

1 Introduction

Turbulent boundary layers under strong adverse pressure gradients are of interest in many technological applications. In the present investigation we focus on the turbulent boundary layer over a flat plate, separating due to a strong adverse pressure gradient. The effect of periodic excitation on the separation bubble is studied by the means of direct numerical simulations. Separated turbulent boundary flow over a flat plate was investigated using direct numerical simulations (DNS) in a few previous studies by [7],[5], [6] and [4], however none of them has studied the effect of periodic excitation on the separation bubble. Periodic excitations have been studied mainly experimentally in several other geometries as a means to influence separated flows. Greenblatt and Wygnanski [1] review the control of flow separation by periodic excitation as well as control and exploitation of airfoil and diffuser flows, including three-dimensional and curvature effects. Important parameters such as the optimum reduced frequencies and excitation levels are discussed.

2 Numerical Method

The simulations of a turbulent boundary layer exposed to an adverse to favourable pressure gradient have been performed using a code developed at the Department of Mechanics by Lundbladh *et al.* [3]. The inflow Reynolds number is $Re_{\delta_0^*} = U_\infty \delta_0^* / \nu = 400$ based on the displacement thickness δ_0^* of the boundary layer and the freestream velocity U_∞ at the inflow $x = 0$. All quantities are non-dimensionalized by U_∞ and δ_0^* at $x = 0$. At this position, a laminar Blasius boundary layer profile is assumed. Downstream at $x = 10$ laminar-turbulent transition is triggered by a random volume force near the wall. The computational box is $450 \times 50 \times 24$. A resolution with 480 modes in streamwise, 193 modes in wall-normal and 64 modes in spanwise direction is used, which gives a total of 6 million points. The boundary conditions are no-slip at the wall. Due to the finite size of the computational domain, the flow domain is truncated and an artificial boundary condition is applied at the freestream.

To generate periodic excitation, we use an oscillating wall-normal body force that exponentially decays from the boundary layer wall and is centred around x_0. We assume the force to be given by

$$F_y = f_0 e^{-y/c} e^{-[(x-x_0)/x_{scale}]^2} \cos(\omega t)\cos(\beta z), \tag{1}$$

where f_0 is the forcing amplitude, ω the oscillation frequency, x_{scale} a parameter controlling the decay of the forcing in x-direction and c a parameter controlling the wall normal decay. The force is causing a wall normal flow. If the parameter $\beta \neq 0$, the force is also varying in spanwise direction.

3 Results

The contours of constant levels of the mean streamwise velocity from -0.025 to 1.025 averaged in time and spanwise direction are shown for the simulation without forcing in Fig. 1b. At the freestream, the streamwise velocity is decelerating between $x = 50$ and $x = 200$ and than accelerating to the value of the freestream velocity at the inlet. This imposes an adverse to favourable pressure gradient on the boundary layer, creating a closed separation bubble. Determining on the zeros of the skin friction coefficient shown in Fig. 1a (*solid line*), the flow separates at $x_{sep} = 126$ and reattaches at $x_{att} = 247$ for the simulation without forcing, so the total reattachment length is $l_{sep} = 121$. The flow can be forced to stay attached in the separated region by applying time-periodic forcing centred around $x_0 = 110$ upstream the separated region with a frequency $\omega = 0.09$, which can be seen from the skin friction coefficient in Fig. 1a (*dashed line*). A close up on the region where the disturbances are generated is shown in Fig. 2a. The isosurfaces of the wall-normal component illustrate the two-dimensional character of the perturbations created by the forcing. These structures decay only slowly while travelling downstream and can be seen travelling upwards in the shear layer. This way, mixing is enhanced. We found that the closer the forcing is placed with respect to the separation point the more effective it is in suppressing separation. It is further observed that sufficiently high amplitudes are necessary in order to reduce the reattachment length, here an amplitude $f_0 = 0.1$ has been used.

However, steady spanwise forcing turned out to be very effective in eliminating the separated region. The skin friction coefficient (Fig. 1 a, *dashed-dotted line*) is affected significantly stronger if forcing is varying in the spanwise direction ($\omega = 0$, $\beta = \frac{2\pi}{24}$) with the same amplitude $f_0 = 0.1$ as in the time-varying case. The effect of the forcing on the flow is illustrated by the isosurfaces of the wall-normal component shown in Fig. 2b. The isosurfaces are displayed

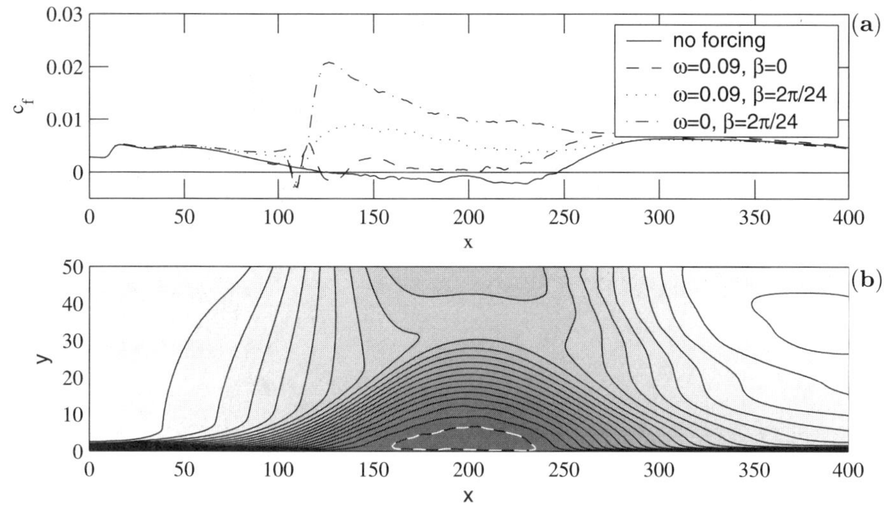

Fig. 1. (**a**) Mean skin friction coefficient, (**b**) Contours of streamwise mean velocity for simulation without forcing (neg: *dark grey* to pos: *white*, *white dashed line*: -0.025)

for a higher value of v in Fig. 2b than in Fig. 2a because of the stronger effect on the flow of the spanwise varying forcing. Even if half the forcing amplitude was used in the case of the spanwise varying forcing, the flow was prevented from separation.

The influence of a forcing which is varying in time and spanwise direction at the same time was also investigated. The value of the skin-friction coefficient for that case is shown in Fig. 1a (*dotted line*) and is located between the curves for the forcing only varying in spanwise direction and for the one only varying in time. Structures of both types of forcing can be seen in Fig. 2c. The fact that spanwise varying forcing showed to be more effective in this flow is probably due to the low Reynolds number in the simulations. At a higher Reynolds number, time-varying forcing is expected to be more effective, since the two-dimensional disturbances are then expected to be more unstable.

A more detailed analysis is subject of the current work considering the linear stability of the flow and the response to spanwise varying forcing.

The work was funded by SAAB and Vinnova through the competence centre PSCI at NADA, KTH.

References

1. D. Greenblatt and I.J. Wygnanski. The control of flow separation by periodic excitation. *Progess in Aerospace Sciences*, 36:487–545, 2000.

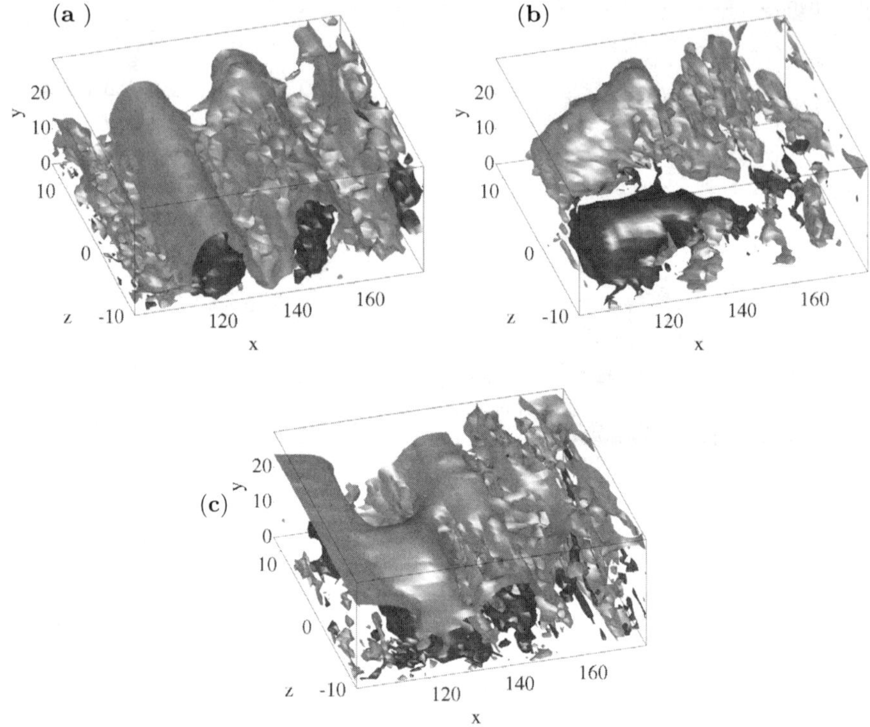

Fig. 2. Instantenous wall-normal velocity component: (**a**) Isosurfaces at values of +0.1 (*light*) and -0.1 (*dark*) for forcing with $\omega = 0.09$, $\beta = 0$ (**b**) Isosurfaces at values of +0.2 (*light*) and -0.2 (*dark*) for forcing with $\omega = 0.0$, $\beta = \frac{2\pi}{24}$ (**c**) Isosurfaces at values of +0.1 (*light*) and -0.1 (*dark*) for forcing at $\omega = 0.09$, $\beta = \frac{2\pi}{24}$.

2. A. H. Herbst and D. S. Henningson. The influence of periodic excitation on a turbulent separation bubble. PSCI Report 03:07, KTH, Stockholm, 2003.
3. A. Lundbladh, S. Berlin, M. Skote, C. Hildings, J. Choi, J. Kim, and D. S. Henningson. An efficient spectral method for simulation of incompressible flow over a flat plate. Technical Report KTH/MEK/TR–99/11–SE, KTH, Department of Mechanics, Stockholm, 1999.
4. M. Manhardt and R. Friedrich. DNS of a turbulent boundary layer with separation. *International Journal of Heat and Fluid Flow*, 471:107–136, 2002.
5. Y. Na and P. Moin. Direct numercial simulation of a separated turbulent boundary layer. *Journal of Fluid Mechanics*, 374:379–405, 1998.
6. M. Skote and D.S. Henningson. Direct numercial simulation of a separated turbulent boundary layer. *Journal of Fluid Mechanics*, 471:107–136, 2002.
7. P.R. Spalart and G.N. Coleman. Numerical simulation of a separation bubble with heat transfer. *Eur. J. Mech. B./Fluids*, 16:169–189, 1998.

Stretching Rate of Passive Lines in Turbulence

Shigeo Kida

Department of Mechanical Engineering, Graduate School of Engineering,
Kyoto University, Yoshida-Honmachi, Sakyo-ku, Kyoto 606-8501, Japan
kida@mech.kyoto-u.ac.jp

The temporal evolution of passive lines in turbulence of an incompressible viscous fluid is simulated numerically for the Taylor-length Reynolds number up to 252. The passive lines elongate in average exponentially in time. The mean exponential stretching rate γ obeys the Kolmogorov scaling law, i.e. $\gamma = 0.17/\tau_\eta$, τ_η being the Kolmogorov time if it is estimated with chopped lines of a fixed length of $O(\mathcal{L})$, $\mathcal{L}$ being the energy-containing-eddy length. However, the mean stretching rate estimated with natural (or unchopped) passive lines increases with Reynolds number more rapidly than at the rate of the Kolmogorov scaling law.

1 Introduction

Strong mixing is one of the most prominent dynamical properites of turbulence. It is now commonly recognized that the coherent vortical structures should play crucial roles in enhancement of mixing, but the precise mechanism is not easy to reveal because of complexity of turbulent motions. We challenge this problem by investigating numerically the motion and the statistics of passive lines in isotropic turbulence to throw light on the mixing phenomena.

It is often said that turbulence is composed of many vortical motions of various shapes and different scales. But the identification of vortices is not an easy task. Recently, an objective eduction scheme of vortices has been proposed and applied to isotropic turbulence [1, 2]. The vortex is identified as a low-pressure swirling region, which takes a long tubular form with high-concentrated vorticity. The shape of the cross-section is different from place to place, but the mean diameter is about 10η independent of the Reynolds number, where η is the Kolmogorov length [3]. Another interesting observation is that two vortices often tend to approach each other in an antiparallel manner [4, 5, 6].

In this paper we show by direct numerical simulation that passive lines in turbulence elongate in average exponentially in time and that the stretching rate obeys the Kolmogorov scaling law if it is estimated for chopped passive lines of a fixed finite length but not for natural (or unchopped) passive lines.

2 Numerical Method

We consider the temporal evolution of passive lines in turbulence, whose velocity field $\boldsymbol{u}(\boldsymbol{x}, t)$ is governed by the Navier-Stokes equation,

$$\left(\frac{\partial}{\partial t} + \boldsymbol{u}(\boldsymbol{x}, t) \cdot \nabla \right) \boldsymbol{u}(\boldsymbol{x}, t) = -\frac{1}{\rho} \nabla p(\boldsymbol{x}, t) + \nu \nabla^2 \boldsymbol{u}(\boldsymbol{x}, t) + \boldsymbol{f}(\boldsymbol{x}, t) \tag{1}$$

and the continuity equation,

$$\nabla \cdot \boldsymbol{u}(\boldsymbol{x}, t) = 0. \tag{2}$$

Periodic boundary conditions are imposed in the three orthogonal directions. Here, ρ is the constant density, $p(\boldsymbol{x}, t)$ is the pressure, ν is the kinematic viscosity, and $\boldsymbol{f}(\boldsymbol{x}, t)$ is an external forcing.

The velocity, the pressure and the forcing fields are expanded into Fourier series. Then, (1) and (2) constitute evolution equations for the Fourier coefficients, which are integrated numerically by the Runge-Kutta-Gill scheme. The nonlinear terms are evaluated by the spectral method dealised by the phase shift algorithm. The amplitudes of Fourier coefficients of velocity in a low-wavenumber range, less than say $\sqrt{8}$, are kept constant in time to realize a statistically stationary state of turbulence. We compare several cases of different values of the Taylor-length Reynolds number,

$$R_\lambda(t) = \frac{U(t)\,\lambda(t)}{\nu} = \sqrt{\frac{20}{3\nu\epsilon(t)}}\,\mathcal{E}(t), \tag{3}$$

ranging between 57 and 252. Here, $U(t)$ is the root mean square of a single component of velocity, $\mathcal{E}(t) = \frac{3}{2}U(t)^2$ is the kinetic energy per unit mass, $\epsilon(t)$ is the energy dissipation rate, and $\lambda(t) = \sqrt{15\nu U(t)^2/\epsilon(t)}$ is the Taylor length.

A passive line is expressed by a set of node points, $\{\boldsymbol{x}_n^{(i)}(t)\}$, where integers n and i indicate the line and node numbers, respectively. Each node $\boldsymbol{x}_n^{(i)}(t)$ is advected by the local velocity as

$$\frac{\mathrm{d}}{\mathrm{d}t}\, \boldsymbol{x}_n^{(i)}(t) = \boldsymbol{u}\big(\boldsymbol{x}_n^{(i)}(t), t\big). \tag{4}$$

The right-hand side of (4) is estimated by the 4^3-point Lagrangian interpolation of $\boldsymbol{u}(\boldsymbol{x}, t)$ at the grid points obtained by the numerical method described above, and the time integration is carried out by the 4th order Runge-Kutta scheme. In order to express a passive line accurately by a set of

Table 1. Statistics of numerical turbulence. Here, τ_η, η and λ are the Kolmogorov time, the Kolmogorov length and the Taylor length, respectively.

Runs	N^3	ν	R_λ	$\mathcal{E}$	ϵ	τ_η	η	λ
7A	128^3	5×10^{-3}	56.5	0.544	0.124	0.202	0.0317	0.469
7B	128^3	2.5×10^{-3}	83.1	0.573	0.127	0.141	0.0187	0.336
8C	256^3	1.25×10^{-3}	121	0.586	0.124	0.101	0.0112	0.243
9D	512^3	6.25×10^{-3}	175	0.585	0.119	0.0726	0.00674	0.175
9E	512^3	3.125×10^{-4}	252	0.601	0.121	0.0509	0.00399	0.125

node points, the distance between any two adjacent nodes must be kept short enough. Whenever the distance exceeds a threshold, 1.5 times the numerical grid width, we add a new node point at a position between the two nodes which is determined by the 4-point Lagrangian interpolation along the line.

3 Stretching Rate of Passive Lines

Numerical simulations with different values of viscosity are performed, and the turbulence statistics are summarized in Table 1. By comparing these six cases, we see that these physical quantities obey the Kolmogorov similarity law, i.e. they vary with viscocity as

$$R_\lambda \propto \nu^{-1/2},\ \mathcal{E} \propto \nu^0,\ \epsilon \propto \nu^0,\ \tau_\eta \propto \nu^{1/2},\ \eta \propto \nu^{3/4},\ \lambda \propto \nu^{1/2}. \tag{5}$$

The total length $L(t)$ of a passive line is calculated by summing up the length of the constituent segments as

$$L(t) = \sum_{i=1}^{I(t)} \ell^{(i)}(t) \qquad \left(\ell^{(i)}(t) = |\boldsymbol{l}^{(i)}(t)|\right), \tag{6}$$

where $\boldsymbol{l}^{(i)}(t) = \boldsymbol{x}_n^{(i+1)}(t) - \boldsymbol{x}_n^{(i)}(t)$, and $I(t)$ is the total number of node points. It is found that $L(t)$ increases in average exponentially in time after an initial transient period.

A quantitative estimation of the stretching rate of passive lines is obtained from the mean value of the exponential stretching rate,

$$\gamma(t) = \frac{\mathrm{d}}{\mathrm{d}t} \log L(t)\,, \tag{7}$$

by taking the average over many passive lines. In order to obtain accurate statistics we track simultaneously M ($= 128^2$ for runs 7A, 7B, 8C, and $= 64^2$ for 9D, 9E) lines in each simulation.

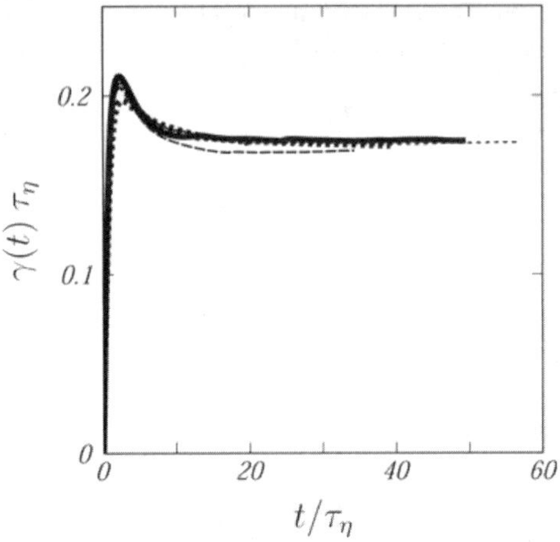

Fig. 1. Temporal evolution of stretching rate of passive lines. The time and the stretching rate are normalized by the Kolmogorov time and its reciprocal, respectively. Thick curve, run 7A; thin dotted, run 7B; thin, run 8C; thin dashed, run 9D; thick dotted, run 9E [7].

3.1 Chopped Passive Lines

Since a passive line elongates in average exponentially in time, the number of node points to describe it may increase soon beyond the capacity of computer memory. In order to suppress the number of node points within the capacity, each passive line is chopped at every time step to keep the length approximately constant of $O(\mathcal{L})$ (see [7] for the algorithm). The temporal evolution of the mean stretching rate calculated by taking the average over M lines and over J $(= 20)$ realizations is plotted in Fig. 1. The time and the stretching rate are normalized by τ_η and ${\tau_\eta}^{-1}$, respectively. The stretching rate starts from zero, peaks around $t \approx 5\tau_\eta$, and settles down to a stationary value after $t \gtrsim 20\tau_\eta$. A constant value of stretching rate implies that the total length of passive lines increases exponentially in time.

The temporal averages of $\gamma(t)$ normalized by ${\tau_\eta}^{-1}$ in the statistically stationary state $(t \gtrsim 20\tau_\eta)$ are listed in Table 2. Since there is no systematic dependence of the average values on the Reynolds number, we may conclude that the time scale of deformation of passive lines is around $(1/0.17 = 5.9)\,\tau_\eta$, which is also comparable to the time scale to forget the initial conditions, i.e., the time scale to settle down to the statistically stationary state (Fig. 1).

3.2 Unchopped Passive Lines

As seen in the preceding subsection the stretching rate of chopped passive lines of a fixed length of $O(\mathcal{L})$ obeys the Kolmogorov scaling law. In order to check whether the stretching rate of natural passive lines obeys the same scaling

Table 2. Stretching rate of chopped passive lines. The mean values of $\gamma\tau_\eta$ are shown for chopped and unchopped passive lines.

Runs	Chopped	Unchopped
7A	0.174	0.184
7B	0.174	0.189
8C	0.175	0.196
9D	0.169	0.222
9E	0.173	0.266

law, we trace the motion of unchopped passive lines. Because of the capacity limit, however, only a single passive line can be simulated in a single run. Then, we take an ensemble average of the stretching rate over 20 independent realizations. The temporal evolution of the stretching rate of passive lines thus obtained is plotted in Fig. 2 for five different values of the Reynolds number. The ordinate and abscissa are normalized as in Fig. 1. It is clearly seen that the stretching rate of unchopped passive lines is larger than that of chopped ones and increases with R_λ.

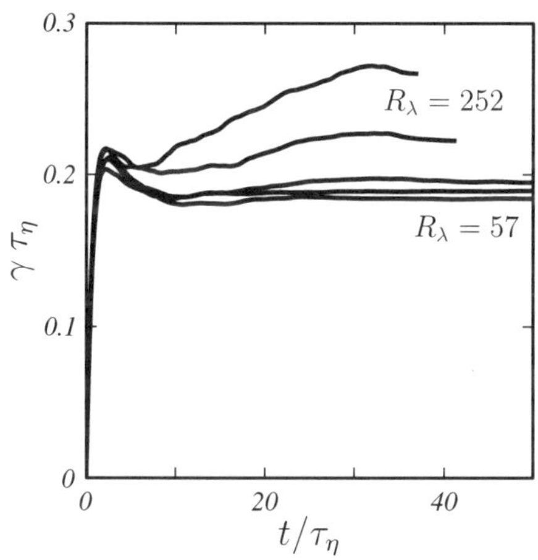

Fig. 2. Stretching rates of unchopped passive lines.

4 Concluding Remarks

We have shown numerically that passive lines in turbulence elongate exponentially in time. The stretching rate calculated by taking an average over many passive lines of a finite length obeys the Kolmogorov scaling law, whereas that of natural passive lines increases with Reynolds number more rapidly than at the rate of the Kolmogorov scaling law. This difference seems to be attributed to the finiteness of the temporal memory of turbulent motions [8, 9]. The portions of passive lines trapped by tubular vortices (mentioned in Introduction) are stretched strongly, which lasts during the trapped time much longer than the Kolmogorov time. A more extensive study is necessary for qualitative confirmation of this theory.

This work has been performed in collaboration with Dr. Susumu Goto under a partial support by a Grant-in-Aid for Science Research on Priority Areas (B) from the Ministry of Education, Culture, Sports, Science and Technology of Japan.

References

1. Kida S, Miura H (1998) Identification and analysis of vortical structures. Eur. J. Mech. B/Fluids 17: 471-488.
2. Kida S, Miura H (1998) Swirl condition in low-pressure vortex. J. Phys. Soc. Jpn. 67: 2166-2169.
3. Makihara T, Kida S (2004) in preparation.
4. Siggia E.D (1985) Collapse and amplification of a vortex filament. Phys. Fluids 28: 794-805.
5. Kida S., Miura H., Adachi T. (2001) Flow strucuture visualization by low-pressure vortex. In: Vassilicos J.C, (ed) Intermittency of turbulent flows. Vol. 69. Cambridge University Press, Cambridge. pp. 262-276.
6. Makihara T, Kida S, Miura H (2002) Automatic tracking of low-pressure vortex. J. Phys. Soc. Jpn. 71: 1622-1625.
7. Goto S, Kida S (2003) Enhanced stretching of material lines by antiparallel vortex pairs in turbulence. Fluid Dyn. Res. 33: 403-431.
8. Kida S, Goto S (2002) Line statistics: stretching rate of passive lines in turbulence. Phys. Fluids 14: 352-361.
9. Goto S, Kida S (2002) A multiplicative process of material line stretching by turbulence. J. Turbulence 3: 17.

Numerical Study of Particle Motion in a Turbulent Ribbed Channel Flow

G. Lo Iacono[1] and P.G. Tucker[2]

[1] Silsoe Research Institute. Wrest Park, Silsoe, Bedfordshire. MK45 4HS, UK. giovanni.loiacono@bbsrc.ac.uk
[2] Civil and Computational Engineering Centre, University of Wales, Swansea, UK P.G.Tucker@swansea.ac.uk

Introduction, Formulation and Methodology

The interaction of spherical and cylindrical particles with a turbulent ribbed channel flow is numerically explored. Large Eddy Simulation (LES) is used to solve the turbulent continuum fluid phase. For the particulate phase, a Lagrangian, one-way coupling, approach is adopted. A simple, potentially highly general, model for cylindrical particles is also proposed, tested and applied to the ribbed channel. Here an LES suitable code [1] is used with Smagorinsky's model. The governing equations are discretised in a centered, second-order, staggered grid, finite-difference framework. The Crank-Nicholson scheme is used to integrate the flow equations in time. A Lagrangian particle trajectory approach with one-way coupling is used. The particles are treated as individual, small, rigid bodies. Spherical and cylindrical particles are injected randomly in the computational domain. In order to calculate trajectories (both for spherical and cylindrical particles), a fourth-order Runge Kutta method is adopted. A sixth-order Lagrangian interpolation is used to evaluate the instantaneous velocity of the fluid $\mathbf{u}_f$ over the particles. For spherical particles, gravity, drag and only when walls are present, Saffman lift forces are included. At low particle Reynolds number, the equation of motion provided by Maxey and Riley [2] is used. Cylinders are decomposed into a series of $i-$segments (in the present case five). Each segment is treated as a cylinder of infinite length, *i.e.* end effects are ignored. The total force acting on each $i-$portion of the cylinder by the surrounding fluid, which is assumed uniform, is decomposed into components perpendicular (D), and parallel (S), to the principal axis of the cylinder $\mathbf{d}$. In addition, D is also parallel to the plane composed by the vectors $\mathbf{d}$ and $(\mathbf{u_f} - \mathbf{u_p})$, where $\mathbf{u_p}$ is the velocity of the particle. The normal component is modelled according the classical drag expression for a cylinder of infinite length [3],

$$D = \frac{1}{2} C_d \rho_f (\mathbf{u_f} - \mathbf{u_p})^2_{\perp} D_p \left\|\mathbf{d}\right\| \tag{1}$$

where $(\mathbf{u_f} - \mathbf{u_p})_\perp$ is the component of the relative velocity perpendicular to $\mathbf{d}$. The other symbols have their usual meanings. The parallel component is modelled as the friction force acting on a flat plate of equivalent surface area to the cylinder, whose module, per unitary surface, is (see for instance [3]):

$$S = \frac{2.586}{Re_l^{0.5}} \frac{1}{2} \rho_f (\mathbf{u_f} - \mathbf{u_p})^2_{\parallel} \tag{2}$$

where Re_l are the Reynolds number based on the length of the cylinder, $(\mathbf{u_f} - \mathbf{u_p})_\parallel$ is the component of the relative velocity parallel to $\mathbf{d}$. Only the component normal to the principal axis generates a torque $\mathbf{T}$. Here $\mathbf{T} = D(\mathbf{d} \times \mathbf{j})$, where $\mathbf{j}$ is the unitary vector in the same direction as D. The components S_i and D_i of all i–elements along the fibre are calculated, leading to the following resultant total forces, $S = \sum S_i$ $D = \sum D_i$ and torque $\mathbf{T} = \sum \mathbf{T}_i$. The velocity $\mathbf{u}_f$ is obtained by interpolating the flow field at the centre of each segment. The total forces D and S cause translation of the cylinder. The torque $\mathbf{T}$ causes rotation around the mass centre. Knowning the forces, the equation of motion (translation and rotation) for a cylinder can be easily evaluated. The extensive LES/particle transport code validation cases (homogeneous and isotropic turbulence, a plane channel flow, and a free shear flow) can be found in [4]. In order to validate basic elements of the cylindrical particle model, some comparisons with analytical results of Cox [9] have been made. For example Fig. 1 plots the dependence of the module of the force acting on the cylinder (the paralled component) *vs* the angle θ between the cylinder and the flow (note, unlike the current model, Cox's model is not valid for $\sin\theta \approx 0$). A similar level of agreement has been obtained for the normal component.

Ribbed Channel

Finally, a LES simulation for a turbulent ribbed channel is made comparing with the $Re = 7000$ measurements of [5]. Periodic boundary conditions are applied in the streamwise (x) and spanwise z directions. Full numerical set-up details can be found in [4]. Simulations, on fine and coarse grids (respectively $215 \times 215 \times 105$ and $113 \times 65 \times 33$ grid points), are carried out. However, it has been observed that particle simulations on these show a similar qualitative behaviour [4]. Therefore, to reduce computational costs and allow more studies, coarse grid results are mainly used. Figure 2 shows the instantaneous velocity field projected on a plane perpendicular to the spanwise direction, in the proximity of the rib. Keeping the same notation of [5], the streamwise distance x (measured from the centre of the rib) and the wall-normal distance y, are expressed in terms of the height of the rib h. Also shown in Fig. 2 are planar weightless particle trajectories. Figure 3 plots the streamwise mean velocity at different downstream locations near the rib, more precisely, $x/h = -0.5$, $x/h = 0.0$, and $x/h = 0.6$. The velocity is normalized by the average flow velocity u_{ref}. The recirculation (i.e. larger negative velocities) downstream of the rib, are captured by the LES. Further downstream of the rib, the LES

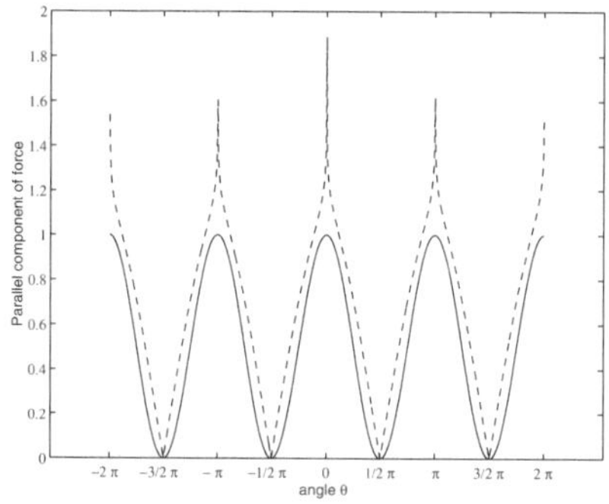

Fig. 1. Dependence of the parallel component *vs* θ. −− Cox's model. −: present model.

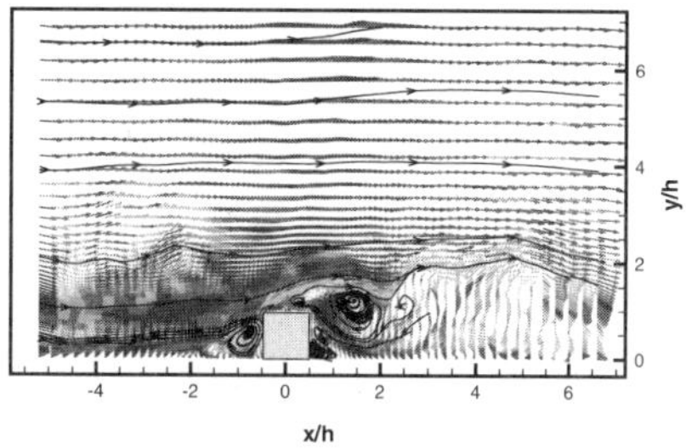

Fig. 2. Instantaneous velocity field and fluid particle motions confined on a plane perpendicular to the spanwise direction. The different gray scale colors correspond to the value of the vorticity magnitude.

profiles have the same trends as the measurements. Cross-stream velocities in the rib vicinity show a similar level of agreement (not shown). The streamwise velocity fluctuations at $x/h = -0.5$, $x/h = 0.0$, and $x/h = 0.6$ can be seen in Fig. 4. The measured $\langle u'^2 \rangle$ exhibits a sharp peak in the near wall rib region. Generally, the finer grid (dashed line) overestimates this peak and coarse (full line) underestimates it. Particle trajectory predictions for the ribbed channel flow are also computed. Among other results, it has been observed, in general agreement with the observations of Wells and Chamberlain [6], that spherical particles tend to accumulate on the rib wall facing the flow direction (see Fig. 5, where the rib position is indicated). The region downstream the rib tends to be particle free. In addition, it is found that particles tend to be trapped in the corner recirculations by the side of the rib. This is suggested in Fig. 6, where the two spikes correspond to the left and right wall of the rib vertical faces. This effect is stronger for lighter particles. Cylindrical particles always avoid the corner recirculations. Instead, except a small percentage that deposit onto the faces of the rib (in this region, the concentration of cylinders is only $\approx 5\%$ of that for spherical particles of the same density), they move towards the central region of the channel between the consecutive ribs.

Conclusions and Recommendations

An LES based particle deposition strategy has been presented. The strategies discussed encompassed the modelling of cylindrical particles using a relatively easy to implement general approach. Essentially, qualitative particle deposition results were then given for a periodic Re $= 7000$ ribbed channel flow. Of course many practical particles are neither perfect spheres nor cylinders. To our knowledge, there is no clear way to model complex particle shapes. A more detailed theoretical analysis for this subject is recommended. The model for cylindrical particles needs more validation. At this stage it is only really truly valid in the limit of an infinitely long cylinder, where no edge effects are present. A more realistic approach ought to model this edge and the effect of the complex flow developed in this region.

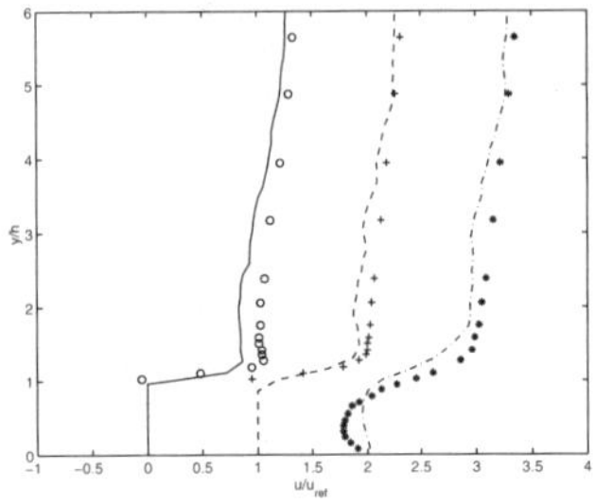

Fig. 3. Streamwise mean velocity at $x/h = -0.5, 0.0, 0.6$. The symbols are the measurements, the lines LES.

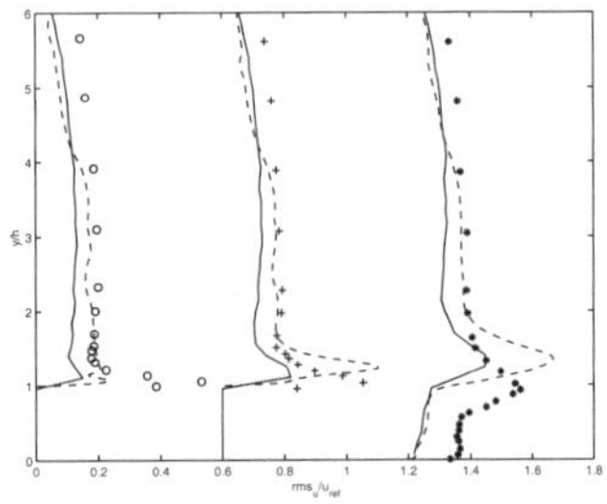

Fig. 4. Streamwise velocity fluctuations at $x/h = -0.5, 0.0, 0.6$. The symbols are the measurements; the continuous lines the current LES; the discontinuous lines an LES on a finer grid.

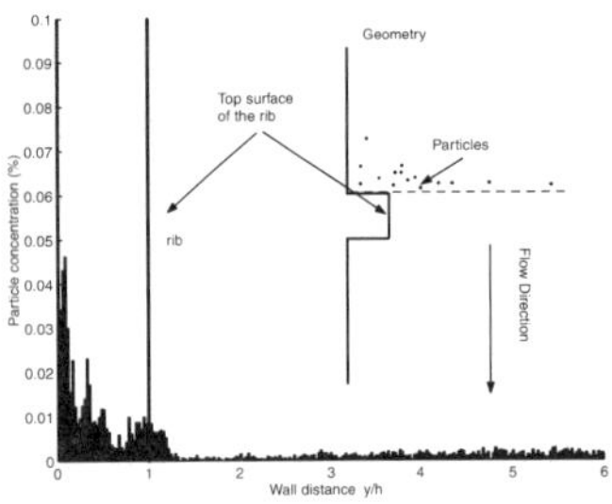

Fig. 5. Geometry and particle concentration at x/h=-0.5.

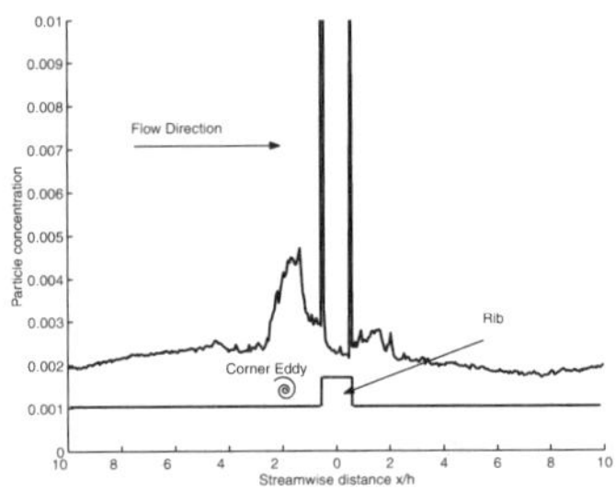

Fig. 6. Geometry and horizontal particle concentration. Light Particles. Details in rib proximity.

References

1. Tucker PG (2001) Computation of unsteady internal flows: fundamental methods with case studies. Kluwer Academic, Norwell Mass
2. Maxey MR. and Riley JJ, (1983) Phys. Fluids, 26:883-889
3. Schlichting H, Gersten K (2000) Boundary-layer theory, Springer, London
4. Lo Iacono G.(2003) Numerical study of particle motion in a turbulent ribbed duct flow. PhD thesis, University of Warwick, United Kingdom
5. Acharya S, Dutta S, Myrum TA, Baker R S (1993) J. Heat Mass Transfer, 36:2069-2982
6. Wells AC, Chamberlain, AC, (1967) Br. J. Appl. Phys., 18:1793-1799

A Fresh Approach to Large Eddy Simulation of Turbulence

R.D. Moser, P. Zandonade, and P. Vedula

University of Illinois at Urbana-Champaign, Urbana IL, 61801, USA
r-moser@uiuc.edu

1 Introduction

One of the most promising techniques for the prediction of turbulent flows is Large Eddy Simulation (LES), in which an under-resolved representation of the turbulence is simulated numerically by modeling the effects of the unresolved small-scales on the simulation. Such simulations have been applied in several flows with reasonable success. However, there are several outstanding problems that need to be addressed before LES can fulfill its promise as a tool for turbulence prediction in engineering flows. The most serious problems limiting the usefulness of LES are the representation of turbulence near walls and other strong inhomogeneities. Other difficulties include the dependence of models on the filter and/or numerical discretization, the treatment of inhomogeneous filters and the lack of understanding of the modeling errors and their impact.

The optimal LES formulation [4, 6, 5, 11, 12] provides a useful framework in which to address these issues and to develop and analyze LES models and simulations. It is this optimal formulation and its application to simple turbulent flow that is the subject of this paper.

1.1 Background on Optimal LES

The optimal LES formulation is reliant on a statistical description of the effects of the small scales. The validity of this statistical perspective depends on an important subtlety regarding filtering. In LES, a spatial filter is commonly used to precisely define the large scales to be simulated. However, there are really two different filtering approaches that can be used in LES [12]:

1. **Continuously Filtered LES:** In this approach, a continuous filter, which maps Navier-Stokes solutions to a space of smoother functions, is used. Examples of such filters are the Gaussian and top-hat filters. Such filters are commonly invertible, or nearly so; that is, the filtered fields contain most if not all of the information required to reconstruct the unfiltered

fields. In practice this is not possible, because the numerical discretization of the field actually discards much of this information. For this reason, the best LES models must depend on both the continuous LES filter and the details of the numerical discretization. Indeed, the above discussion suggests that the discretization may be more important, which leads to consideration of the discrete filtering approach.

2. **Discretely Filtered LES:** In the discrete filter approach, the filter is considered to be a mapping from the infinite-dimensional space in which Navier-Stokes solutions reside to a finite-dimensional space which can be represented numerically without further discretization. This avoids the problems in continuously filtered LES by combining the filtering and discretization into a single non-invertible linear mapping. The mapping is clearly many-to-one, so there are many (formally infinite) fields which all map to the same filtered field. In the optimal LES approach, we consider a statistical description of the evolution of all these fields.

An important mathematical result [4], which motivates the optimal LES approach, is that an LES w will match the one-time statistics of filtered turbulence $\tilde{u}$ if and only if the model $m_i(w)$ of the subgrid model term M_i is given by

$$m_i(w) = \langle M_i(u) | \tilde{u} = w \rangle \tag{1}$$

This model also minimizes the difference between M_i and m_i (in the mean-square sense), and so this model has all the properties that one could ask of a sub-grid model. We therefore call it the ideal sub-grid model.

Unfortunately, the conditional average in (1) cannot practically be determined, since the conditions are that the entire filtered velocity field match the entire LES field. However, it can be estimated using stochastic estimation [2] which is a well-established technique for estimating conditional averages. Optimal LES is essentially the use of stochastic estimation for LES modeling, which is an approach first proposed by Adrian [1].

Even stochastic estimation requires a large amount of statistical information. Until recently, this has primarily come from direct numerical simulations and experiments. However, for LES applications, this need for extensive empirical data is unacceptable. The primary focus of recent efforts has been the theory and modeling needed to replace this detailed data. Because practical applications of LES are generally in complex geometries for which finite volume discretizations are commonly used, much of the effort is aimed at LES representations based on these discretizations.

In the remainder of the paper, some Optimal LES results based on DNS statistical data are presented as a validation of the optimal modeling approach. Then, an approach for the determination of optimal models without the use of extensive data is presented, along with some preliminary results for models developed in this way. Finally, the prospect for developing practical LES models using the Optimal LES approach are discussed.

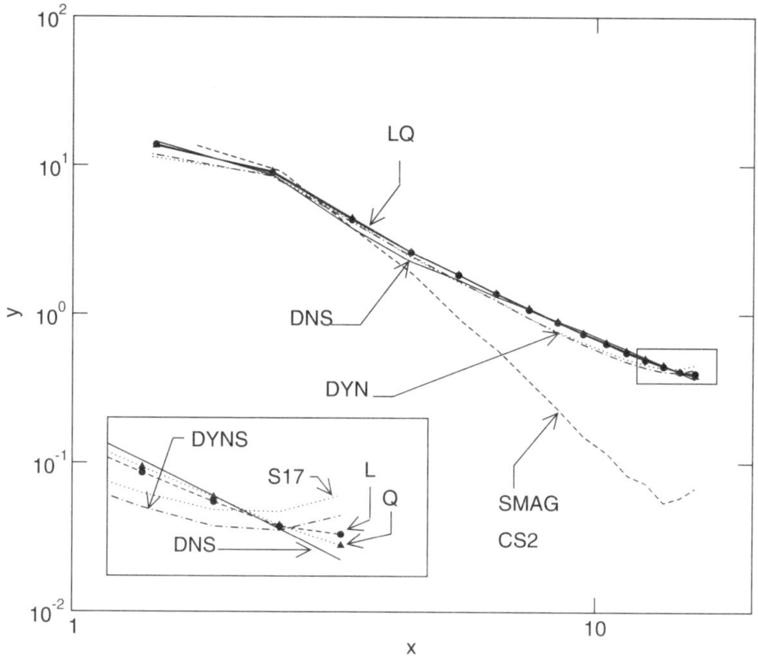

Fig. 1. The three dimensional energy spectrum, $E(k)$, for filtered DNS and 32^3 LES with an optimal linear model (L^{16}), an optimal quadratic model (Q^{16}), two cases of a Smagorinsky model ($Cs = 0.17$ and $Cs = 0.819$) and the Dynamic Smagorinsky model (Dynamic).

2 Results Based on DNS Statistical DATA

To validate the Optimal LES approach, models were constructed and LES were performed for three different situations: 1) isotropic turbulence with Fourier-cutoff filters, 2) turbulent channel flow with Fourier cutoff filters in the directions parallel to the wall, and 3) isotropic turbulence with "finite-volume" filters. In this section, example results from these studies are presented.

For isotropic turbulence, DNS data were obtained for a forced isotropic turbulence, with forcing in the lowest three modes [4]. The micro-scale Reynolds number was $Re_\lambda = 164$, and the DNS were performed with 256^3 points using a Fourier spectral method like that of Rogallo [10]. A sharp Fourier cutoff filter with cutoff wavenumber $k_c = 16$ was used, to produce a 32^3 LES. Both linear and quadratic Optimal LES models were developed using the DNS statistical data, and were tested. In figure 1 the three-dimensional spectrum for the DNS, the Optimal LES and several other models are shown. It is clear that the Optimal LES are in excellent agreement with the DNS, better than for the other models.

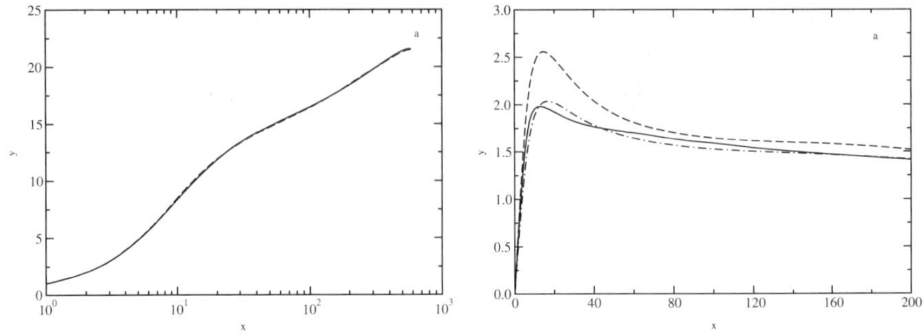

Fig. 2. Mean and rms velocity profiles in wall coordinates from Optimal LES based on estimating subgrid stress in terms of velocity and first two wall-normal derivatives (———), DNS (– – –) filtered DNS (- · - · -). Shown are (a) mean velocity, (b) rms streamwise velocity u_{rms}.

In turbulent channel flow, DNS data from the $Re_\tau = 590$ simulation of Moser *et al* [8] were used to develop optimal LES models. In this case, the DNS representation used 384 Fourier spectral modes in the streamwise and spanwise directions. Fourier cutoff filters in these directions were used to define the LES, reducing the number of modes in each direction to 32. The effective filter width in the spanwise and streamwise directions is then 58 and 116 wall units respectively. There was no filtering in the wall-normal direction [11]. Linear optimal LES models were developed using the DNS statistical data, but due to the inhomogeneity, the dependence of the model on the filtered velocities was local in the wall-normal direction. Several optimal models were developed which differed in the form of the model term (either the divergence of subgrid stress or the stress itself) and in the model dependence (dependence on wall-normal derivatives). Results for the mean and streamwise rms velocity profiles are shown in figure 2 for one of the cases. See Volker *et al* [11] for results from all the cases studied.

In this case the subgrid stress τ_{ij} is modeled in terms of the velocities and their first two wall-normal derivatives. Notice that the mean velocity almost exactly matches the DNS, while the rms velocity is in reasonable agreement with the profiles for the filtered DNS. Results for the other cases (not shown, see Volker [11]) are much worse than for that shown here. The reason for this difference is that by optimally modeling of τ_{ij}, rather than its divergence or other terms, the optimal model is constructed to correctly reproduce the subgrid contribution to the wall-normal transport of resolved Reynolds stress ($\partial u_2 \tau_{ij}/\partial x_2$), when measured *a priori* [11]. In essence, the optimal model for τ_{ij} represents transport as well as transfer of energy and Reynolds stress to small scales, while optimal modeling $\partial \tau_{ij}/\partial x_j$ only represent transfer to small scales.

Finally, the isotropic turbulence case is repeated, only this time the filter is defined through averaging over finite volumes. The motivation for testing this filter is that finite volume methods are much more appropriate for use in

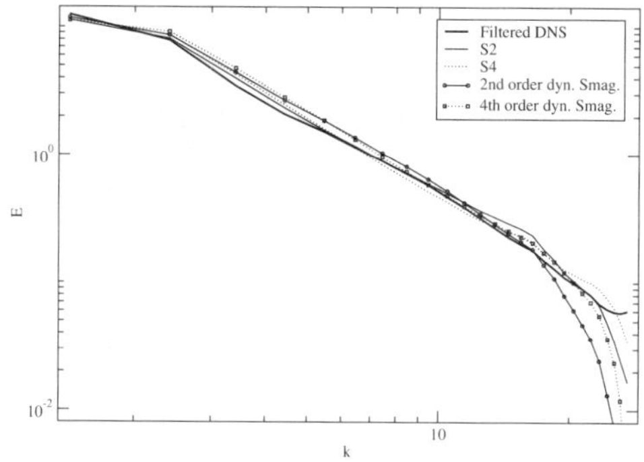

Fig. 3. Three-dimensional energy spectrum E(k), filtered DNS compared with *optimal* LES with 2-cell (S2) and 4-cell (S4) stencils and dynamic Smagorinsky.

general geometries. Just as in a finite volume numerical method, the application of the filter to the Navier-Stokes equations yields an expression for the time-derivative of the volume-averaged velocity in terms of the momentum fluxes through the volume faces. However, in the LES case, the volumes are large compared to the small scales of turbulence, so standard finite volume reconstructions cannot be used to determine the fluxes. Instead, we will use optimal LES techniques to estimate the fluxes in terms of average velocities in near-by volumes [12]. In this approach, there is no distinction between the numerical representation of the Navier-Stokes equations, and the subgrid model. Further, the estimation dependencies must at least include quadratics to be able to represent the convective term.

The LES finite volume grid was 32^3, and the correlations required for the optimal LES formulation were determined from the isotropic DNS simulations described above. Various "stencils" defining the dependencies on velocity components and neighboring volumes were used in the estimates. The results shown in figure 2 indicate that for sufficiently large stencils (4-cells, 2 cells on either side of each face), the LES results are in reasonable agreement with the filtered DNS. In particular, the 4-cell stencil yields results superior to the dynamic Smagorinsky model.

The validation test results described above indicate that the Optimal LES approach to developing LES models yields remarkably good LES results, when the models are appropriately constructed. In particular, it appears that for strongly inhomogeneous flows, it is necessary for the model to correctly represent the subgrid contribution to terms in the Reynolds stress equations, such as transport. The optimal modeling approach is well suited to constructing models that do this [11].

3 Theory-Based Optimal Models

While the DNS-based Optimal LES described above was very helpful in validating the general modeling approach, if such models are to be useful for practical simulations, the models must not depend on extensive empirical input, such as the DNS correlations. Instead, theoretical considerations are used to determine the correlations that go into the models. Here, the finite volume filtering approach is used, because it is the most useful in practical simulations. In the finite volume formulation, with quadratic dependencies, the equations defining the model are:

$$M'_{ij} = \sum_{\alpha} L^{\alpha}_{ijk} w^{\alpha}_k + \sum_{\alpha,\beta} Q^{\alpha\beta}_{ijkl} (w^{\alpha}_k w^{\beta}_l)' \tag{2}$$

$$\langle w^{\gamma}_m M'_{ij} \rangle = \sum_{\alpha} L^{\alpha}_{ijk} \langle w^{\alpha}_k w^{\gamma}_m \rangle + \sum_{\alpha,\beta} Q^{\alpha\beta}_{ijkl} \langle (w^{\alpha}_k w^{\beta}_l)' w^{\gamma}_m \rangle \tag{3}$$

$$\begin{aligned} \langle (w^{\gamma}_m w^{\delta}_n)' M'_{ij} \rangle = & \sum_{\alpha} L^{\alpha}_{ijk} \langle w^{\alpha}_k (w^{\gamma}_m w^{\delta}_n)' \rangle \\ & + \sum_{\alpha,\beta} Q^{\alpha\beta}_{ijkl} \langle (w^{\alpha}_k w^{\beta}_l)' (w^{\gamma}_m w^{\delta}_n)' \rangle \end{aligned} \tag{4}$$

where M_{ij} is the surface flux of momentum component i through face j, w^{α}_i is velocity component i averaged over volume α, L^{α}_{ijk} and $Q^{\alpha\beta}_{ijkl}$ are the estimation coefficients and primes represent fluctuating components.

The inputs to the model are the multi-point correlations ($< \cdot >$ represents expected value) among the LES variables w_j on the right, and the correlations of w_j with the flux term M_i. Equations (3) and (4) must be solved for the coefficients L_{ijk} and Q_{ijkl}. Correlations on the right-hand side among the w_j are correlations among the LES state variables, and can in principle be determined from a running LES. This is particularly important for the triple correlations because the theory described below provides no guidance for them. In contrast, the correlations on the left involve the unknown flux terms, so they must be specified.

The flux correlations on the left are surface/volume integrals of the two-point third-order velocity correlations and the three-point fourth-order velocity correlations. Thus, we only need theoretical results for these quantities. If we assume that the Reynolds number is very large (particularly the cell Reynolds number) and that the separations (i.e. the volume sizes) are in the inertial range, it is justified to use the Kolmogorov expressions for these correlations (the 2/3 law and the 4/5 law), as well as small-scale isotropy. Further, despite difficulties in using the quasi-normal approximation in other contexts, the quasi-normal approximation can be used for the fourth-order correlations. Because we are not using the quasi-normal approximation to close statistical equations, such problems are not expected. With these assumptions, the following expressions for the required correlations can be determined:

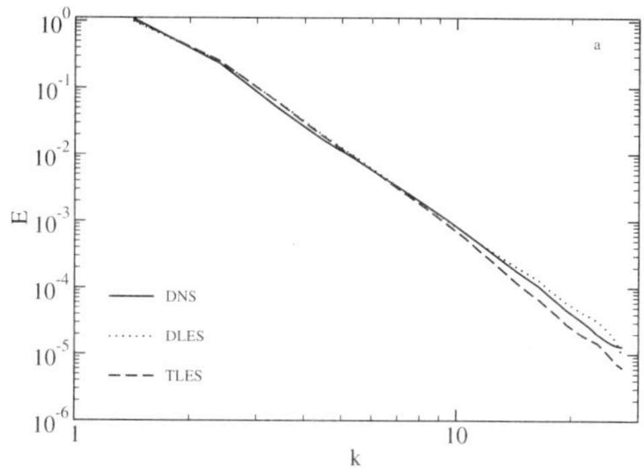

Fig. 4. Three-dimensional energy spectrum E(k), filtered DNS compared with *Optimal* LES with 4-cell stencils, developed from theory and from DNS statistical data.

$$\langle u_i(\mathbf{x})u_j(\mathbf{x}')\rangle = u^2\delta_{ij} + \frac{C_1}{6}\epsilon^{2/3}r^{-4/3}(r_ir_j - 4r^2\delta_{ij}) \quad (5)$$

$$\langle u_i(\mathbf{x})u_j(\mathbf{x})u_m(\mathbf{x}')\rangle = \frac{\epsilon}{15}\left(\delta_{ij}r_m - \tfrac{3}{2}(\delta_{jm}r_i + \delta_{im}r_j)\right) \quad (6)$$

$$\begin{aligned}\langle u_i(\mathbf{x})u_j(\mathbf{x})u_k(\mathbf{x}')u_l(\mathbf{x}'')\rangle &= \langle u_i(\mathbf{x})u_j(\mathbf{x})\rangle\langle u_k(\mathbf{x}')u_l(\mathbf{x}'')\rangle \\ &+ \langle u_i(\mathbf{x})u_k(\mathbf{x}')\rangle\langle u_j(\mathbf{x})u_l(\mathbf{x}'')\rangle \\ &+ \langle u_i(\mathbf{x})u_l(\mathbf{x}'')\rangle\langle u_k(\mathbf{x}')u_l(\mathbf{x})\rangle \end{aligned} \quad (7)$$

By integrating these expressions over appropriate volumes and faces, the correlations required in (3-4) can be computed.

This theoretical approach was used in the isotropic turbulence case to determine all the correlations appearing in (3-4), except the third-order three-point correlations that appears on the right of these equations. These three-point third-order correlations were determined from the isotropic DNS described above, because no theoretical expression for them is available. The LES results using these theoretical models for the 4-cell stencil are shown in figure 4 along with the DNS-based LES and the filtered DNS. The spectrum from the theoretically-based LES is in somewhat poorer agreement with the filtered DNS, but there is a slight inconsistency due to the use of the DNS data for the third-order correlations, correcting this may improve the results.

The results from the theory-based Optimal LES described above are encouraging; however, there are a number of the improvements that are required before these models can be deployed for use in practical simulations. First, a dynamic procedure to determine the correlations on the right hand sides of (3-4) from the running DNS is required, to eliminate the need for the last bit of DNS data. Such a dynamic approach is currently being tested. Second, the theory described above assumes that the small scales are approximately homogeneous and isotropic. This will need to be generalized for use with strongly inhomogeneous flows, such as near walls. An effort is currently underway to characterize the near-wall correlations that are needed in terms of a few empirically determined coefficients. A variety of theoretical approaches are available for this. These include: the anisotropy representations of Arad *et al* [3] and L'vov *et al* [7], the log-layer similarity representation of Oberlack [9], consistency with the two-point correlation equation, and the quasi-normal approximation.

4 Conclusions

Optimal LES modeling [4, 5, 11, 12] yields remarkably good LES results. The strengths of the approach are: 1) it provides a technique to design models to represent important statistical properties; 2) it is perfectly general, applying equally well near walls and to turbulence away from walls; and 3) it provides a framework in which our knowledge of turbulence can be applied to LES modeling. However, substantial multi-point correlation inputs are required. It was shown that these inputs can be provided theoretically for isotropic inertial-range small scales. However, near walls and other strong inhomogeneities, other considerations are required. The correlations needed in the Optimal LES modeling of near-wall turbulence are being developed.

References

1. R. Adrian. Stochastic estimation of sub-grid scale motions. *Applied Mechanics Review*, 43(5):214–218, 1990.
2. R. Adrian, B. Jones, M. Chung, Y. Hassan, C. Nithianandan, and A. Tung. Approximation of turbulent conditional averages by stochastic estimation. *Physics of Fluids*, 1(6):992–998, 1989.
3. I. Arad, V.S. L'vov, and I. Procaccia. Anomalous scaling in anisotropic turbulence. *Physica A*, 288:280–307, 2000.
4. J. Langford and R. Moser. Optimal LES formulations for isotropic turbulence. *Journal of Fluid Mechanics*, 398:321–346, 1999.
5. J. A. Langford and R. D. Moser. Breakdown of continuity in large-eddy simulation. *Phys. of Fluids*, 11:943–945, 2001.
6. J.A. Langford. *Toward Ideal Large-Eddy Simulation*. PhD thesis, University of Illinois at Urbana-Champaign, 2000.
7. V. S. L'vov, I. Procaccia, and V. Tiberkevich. Scaling exponents in anisotropic hydrodynamic turbulence. *Phys. Rev. E*, 67:026312, 2003.
8. R.D. Moser, J. Kim, and N.N. Mansour. Direct numerical simulation of turbulent channel flow up to $Re_\tau = 590$. *Physics of Fluids*, 11(4):943–945, April 1999.
9. M Oberlack. A unified approach for symmetries in plane parallel turbulent shear flows. *Journal of Fluid Mechanics*, 427:299–328, 2001.
10. R. Rogallo. Numerical experiments in homogeneous turbulence. Technical Report TM-81315, NASA Ames, 1981.
11. S. Volker, P. Venugopal, and R. D. Moser. Optimal large eddy simulation of turbulent channel flow based on direct numerical simulation statistical data. *Phys. of Fluids*, 14:3675, 2002.
12. P. S. Zandonade, J. A. Langford, and R. D. Moser. Finite-volume optimal large-eddy simulation of isotropic turbulence. To appear in *Phys. of Fluids*, 2004.

Statistical Analysis of Turbulent Natural Convection in Low Prandtl Number Fluids

I. Otić and G. Grötzbach

Forschungszentrum Karlsruhe GmbH, Institut für Kern- und Energietechnik
otic@iket.fzk.de, groetzbach@iket.fzk.de

Summary. Direct numerical simulations (DNS) of turbulent Rayleigh-Bénard convection for a liquid metal with $Pr = 0.025$ are used to perform statistical analysis, in particular to evaluate the turbulent diffusion term in the temperature variance equation. These results are compered with DNS based predictions by a standard model of the turbulent diffusion term in the temperature variance equation and by a model which was recently developed by the authors.

1 Introduction

Statistical turbulence closures serve currently as a basis of applied computational fluid dynamics (CFD). Common turbulent heat flux models are based on the Reynolds analogy. Due to the large thermal and small viscous diffusivity of low Prandtl number fluids, like liquid metals, the statistical properties of the momentum and the temperature fields are non-similar, so that the turbulent heat flux models based on the Reynolds analogy cannot be applied (see e.g. [6]). The development of improved statistical heat flux models needs input from measurements, but because of some specific properties of these fluids (like aggressive chemistry) we have still no sensors to measure accurately small scale high frequency fluctuations of the velocity and of the cross-correlations between velocity and temperature fluctuations. DNS is a tool which can provide such data to improve and to develop improved statistical turbulent heat flux models as it was shown by realizations in a CFD code [1].

In this paper, results of the direct numerical simulation of the Rayleigh-Bénard convection for the Prandtl number $Pr = 0.025$ (lead-bismuth) and Rayleigh number $Ra = 100,000$ are used to perform some statistical analysis of turbulence in this convection type. The results are also used to compare the performance of a standard and of a recently developed statistical turbulence model for the turbulent diffusion of the temperature variance [6] with the DNS results.

2 Physical and Numerical Model

2.1 Rayleigh-Bénard Convection

A physical model for the investigation of heat transfer by natural convection is the Rayleigh-Bénard convection. It is given by an infinite fluid layer which is confined by two rigid horizontal isothermal walls. The lower one is heated and the upper one is cooled. The physical problem is characterized by following dimensionless numbers: The Rayleigh number $Ra = \frac{g\beta \Delta T D^3}{\nu\kappa}$; the Prandtl number $Pr = \frac{\nu}{\kappa}$; and the Grashof number $Gr = \frac{Ra}{Pr}$; where g is the gravity; β is the thermal expansion coefficient; ΔT is the wall temperature difference; D is the distance between the two horizontal walls; ν is the kinematic viscosity; and κ is the thermal diffusivity.

2.2 Direct Numerical Simulation of the Turbulent Rayleigh-Bénard Convection

Direct numerical simulation is a method in which the three dimensional time dependant conservation equations for mass, momentum and energy are solved numerically such that all relevant physical processes are resolved by the grid and by the computational domain. For an incompressible Newtonian fluid with constant material properties the conservation equations can be written in the following dimensionless form:

$$\frac{\partial u_i}{\partial x_i} = 0, \tag{1}$$

$$\frac{\partial u_i}{\partial t} + \frac{\partial (u_i u_j)}{\partial x_j} = -\frac{\partial p}{\partial x_i} + \frac{1}{\sqrt{Gr}} \frac{\partial}{\partial x_j} \left(\frac{\partial u_i}{\partial x_j} + \frac{\partial u_j}{\partial x_i} \right) - (T_{ref} - T)\delta_{i3}, \tag{2}$$

$$\frac{\partial T}{\partial t} + \frac{\partial (T u_j)}{\partial x_j} = \frac{1}{\sqrt{PrRa}} \frac{\partial^2 T}{\partial x_j \partial x_j}, \tag{3}$$

where T_{ref} is a reference temperature and δ_{ij} is the Kronecker delta. Equations (1)-(3) are normalized by the channel height D, velocity $u_0 = \sqrt{g\beta \Delta T D}$, time D/u_0, pressure ρu_0^2, and difference between the temperatures of the two walls ΔT. Simulations of Rayleigh-Bénard convection are performed with the TURBIT code (see [3]). It is a finite volume code which allows for direct numerical simulations of turbulent heat and mass transfer in simple channel geometries. The boundary conditions use periodicity in both horizontal directions, whereas the no slip condition and constant wall temperatures are specified at the lower and upper wall. Here X_1, X_2 assign the horizontal and X_3 the vertical direction.

3 Statistical Analysis

In the following we assign a time averaged quantity with $\overline{X}$. Numerically $\overline{X}$ is determined by averaging the data over both homogeneous horizontal directions and over time. Here we denote with $U = \overline{U} + u$, and $T = \overline{T} + \theta$ the Reynolds decomposition of velocity and temperature.

Fig. 1 shows the mean temperature profile evaluated from the DNS results. We find very thick thermal boundary layers, so convection has a small influence on the heat transfer. Therefore, the Rayleigh number is obviously not large enough to build up an area with a constant temperature in the middle of the channel. The temperature field is equallly influenced by both, by conduction and by convection.

Modeling of the turbulent heat fluxes which is not based on Reynolds analogy usually considers the temperature variance $\overline{\theta^2}$ (see e.g. [5]). A closure of the transport equation for the temperature variance requires a model for the turbulent diffusion $D_\theta^t = \frac{\partial}{\partial x_i}\overline{u_i\theta^2}$ (see e.g. [6]).

The vector $\overline{u_i\theta^2}$ is usually modeled using a generalized gradient diffusion hypothesis (GGDH), first introduced by Daly and Harlow [2], as

$$\overline{u_i\theta^2} = -C_{D\theta}\frac{k}{\epsilon}\overline{u_iu_j}\frac{\partial\overline{\theta^2}}{\partial x_j}, \tag{4}$$

where the standard coefficient $C_{D\theta} = 0.22$ after Jones and Musonge [4] is used. Recently a new model for the triple correlations $\overline{u_i\theta^2}$ is derived in [6] resulting in:

$$\overline{u_i\theta^2} = -C_\theta\sqrt{\frac{k}{\epsilon}\frac{\overline{\theta^2}}{\epsilon_\theta}}\left[2\sqrt{\nu\kappa}\Delta_x\overline{u_i\theta^2} + \overline{u_iu_j}\frac{\partial\overline{\theta^2}}{\partial x_j}\right], \tag{5}$$

where the coefficient $C_\theta = 0.11$ is used.

Fig. 2 shows a comparison of the DNS results for the turbulent diffusion D_θ^t and of the DNS based predictions by the models (4) and (5). The standard model (eq. (4)) considers only the mechanical time scale $\frac{k}{\epsilon}$. This modeling approach obviously does not account for the molecular fluid properties, resulting in strong over-predictions of the DNS results. The model (5) accounts for the molecular fluid properties and considers a mixed time scale $\frac{k}{\epsilon}\frac{\overline{\theta^2}}{\epsilon_\theta}$. This model reproduces the DNS data very well.

4 Conclusion

Results of the direct numerical simulations of turbulent Rayleigh-Bénard convection for $Pr = 0.025$, $Ra = 10^5$ are used to show that at these Rayleigh and Prandtl numbers the temperature field is still influenced by conduction. Results of statistical analysis are used to evaluate the turbulent diffusion term in the temperature variance equation. These results are compered with DNS

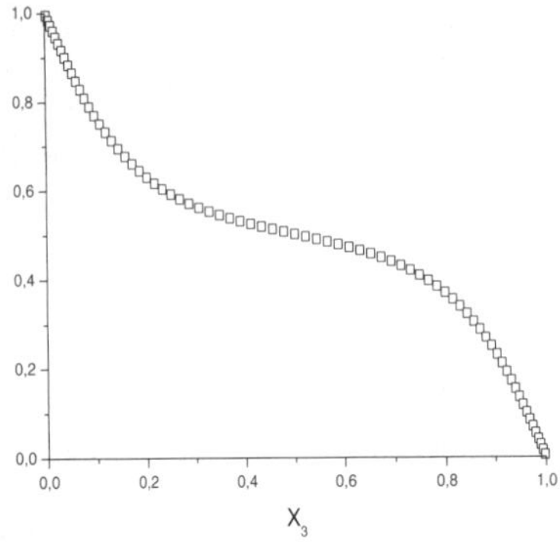

Fig. 1. Mean temperature profile for $Pr = 0.025$, $Ra = 100,000$

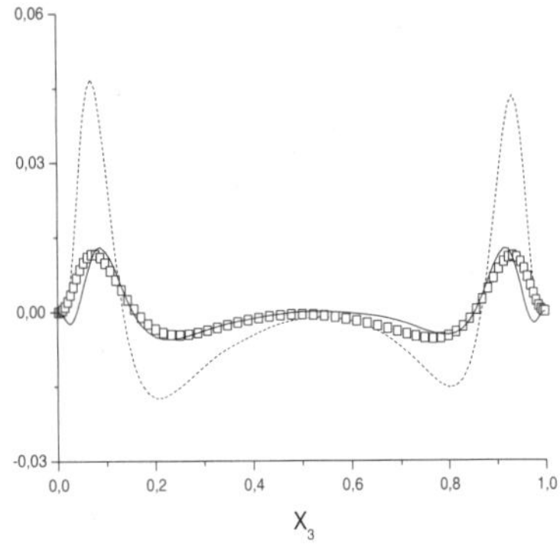

Fig. 2. DNS evaluated vertical profile of D_θ^t □; Predictions by the model (4) $\cdots$ and by the model (5) —

based predictions by a standard and by a recently developed model which accounts for the molecular fluid properties and therefor should be applicable for a wide range of Prandtl numbers.

This research received financial support (GR 1901) from the Deutsche Forschungsgemeinschaft (DFG). The authors gratefully acknowledge this support.

References

1. L.N. Carteciano and G. Grötzbach. Validation of turbulence models in the computer code FLUTAN for a free hot sodium jet in different buoyancy flow regimes. Technical Report FZKA 6600, Forschungszentrum Karlsruhe, 2003.
2. B.J. Daly and F.H. Harlow. Transport equations in turbulence. *Phys. Fluids*, 13:2634–2649, 1970.
3. G. Grötzbach. Direct numerical and large eddy simulation of turbulent channel flows. In N.P. Cheremisinoff, editor, *Encyclopaedia of fluid mechanics*, pages 1337–1391. Gulf Publ. Houston, 1987.
4. W.P. Jones and P. Musonge. Closure of the Reynolds-stress and scalar flux equations. *Phys. Fluids*, 31:3598–3604, 1988.
5. B.E. Launder. On the computation of convective heat transfer in complex turbulent flows. *Trans. ASME*, 110:1112–1128, 1988.
6. I. Otić, G. Grötzbach, and M. Wörner. Analysis and modelling of the temperature variance equation in turbulent natural convection for low Prandtl number fluids. *submited for publication*, 2004.

Computational Simulation of Transitional and Turbulent Shear Flows

P. Schlatter, S. Stolz, and L. Kleiser

Institute of Fluid Dynamics, ETH Zürich, CH-8092 Zürich, Switzerland
schlatter@ifd.mavt.ethz.ch

A short review of numerical simulation approaches for transitional and turbulent shear flows is presented. Some results using large-eddy simulation (LES) are for canonical turbulent and transitional flows obtained with different subgrid-scale (SGS) models such as a variant of the approximate deconvolution (ADM) and high-pass-filtered (HPF) eddy-viscosity model. Special focus is the LES of transition in incompressible flow.

1 Introduction

The demand for turbulent flow computations has been increasing strongly in the recent past, driven by the need for faster and more accurate flow predictions in numerous applications and by wide-spread access to exponentially growing computing power.

The most accurate approach to solve turbulent flow problems is the direct numerical simulation (DNS), in which all relevant length and time scales of the flow are resolved on the numerical grid [9]. The required computation time typically increases with the third power of the Reynolds number Re of the flow, which severely limits the range of problems accessible to DNS.

Practical high Reynolds-number calculations thus need to be performed using simplified turbulence models. Conventionally, numerical computations of engineering or geophysical turbulent flows are based on the Reynolds-averaged Navier-Stokes (RANS) equations for which statistical turbulence models must be chosen. Only statistical information about turbulence can be obtained by such computations. Although this technique may require a number of *ad-hoc* adjustments of the turbulence model to a particular flow situation, quite satisfactory results can be obtained in many practical cases.

A technique with a level of generality in between DNS and RANS is the large-eddy simulation (LES). The eddies (turbulent vortices) above a certain size are completely resolved on the numerical grid, whereas the smaller scales and their interaction with the resolved scales are modeled. The idea behind this scale-separation is that the smaller eddies are more homogeneous and isotropic and depend little on the specific flow situation, whereas the energy-carrying large-scale vortices are strongly affected by the particular flow conditions, e.g. geometry, inflow, etc. For a LES, only a fraction of the computational cost compared to DNS is needed (typically of order 1%). The success of LES is essentially dependent on the quality of the underlying sub-grid scale (SGS) model and the applied numerical solution scheme. Substantial research efforts during the past 20 years have led to more universal SGS models, see e.g. [6, 1].

It is expected that LES will play a major role in the future for prediction and analysis of certain complex turbulent flows in which a representation of unsteady turbulent fluctuations is important, such as large-scale flow separation in aerodynamics, coupled fluid-structure interaction, turbulent flow control, aeroacoustics and turbulent combustion.

2 Large-Eddy Simulation

The governing equations for transitional and turbulent flows are the Navier-Stokes equations for the velocity components u_i $(i = 1, 2, 3)$ and the pressure p, given here for an incompressible flow complemented with the incompressibility constraint,

$$\frac{\partial u_i}{\partial t} + \frac{\partial u_j u_i}{\partial x_j} = -\frac{\partial p}{\partial x_i} + \frac{1}{Re}\frac{\partial^2 u_i}{\partial x_j \partial x_j} \qquad \text{and} \qquad \frac{\partial u_i}{\partial x_i} = 0 \; . \tag{1}$$

In the traditional LES approach, these equations are spatially filtered by a low-pass filter G with a filter width Δ. The filter operation yields the LES equations

$$\frac{\partial \overline{u}_i}{\partial t} + \frac{\partial \overline{u}_j \overline{u}_i}{\partial x_j} = -\frac{\partial \overline{p}}{\partial x_i} - \frac{\partial \tau_{ij}}{\partial x_j} + \frac{1}{Re}\frac{\partial^2 \overline{u}_i}{\partial x_j \partial x_j} \qquad \text{and} \qquad \frac{\partial \overline{u}_i}{\partial x_i} = 0 \; . \tag{2}$$

The low-pass filtered velocity is given by

$$\overline{u}_i(x) := G * u_i := \int_{\mathcal{V}} G(x, x', \Delta) u_i(x') dx' \; , \tag{3}$$

where $\mathcal{V}$ is the computational domain. The LES equations govern the evolution of the large, energy-carrying scales of motion. The effect of the non-resolved small scales enters through a subgrid-scale (SGS) term

$$\tau_{ij} = \overline{u_i u_j} - \overline{u}_i \overline{u}_j \tag{4}$$

which is not closed (i.e. cannot be obtained from the computed quantities $\overline{u}_i$) and must thus be modeled by a SGS model.

The most widely used SGS models are the eddy-viscosity models, given by

$$\tau_{ij} - \frac{\delta_{ij}}{3}\tau_{kk} = -2\nu_T \overline{S}_{ij} , \tag{5}$$

where $\overline{S}_{ij}$ is the large-scale strain-rate tensor $\overline{S}_{ij} = (\partial\overline{u}_i/\partial x_j + \partial\overline{u}_j/\partial x_i)/2$. The eddy-viscosity ν_T is usually modeled according to [14] yielding the widely-used Smagorinsky model

$$\nu_T = (C_S \Delta)^2 |\overline{S}| \quad , \quad |\overline{S}| = (2\overline{S}_{ij}\overline{S}_{ij})^{1/2} . \tag{6}$$

The model coefficient C_S ("Smagorinsky constant") has to be determined empirically and is dependent on the particular flow situation.

Classical variants of eddy-viscosity models with constant model coefficients, e.g. the above-mentioned Smagorinsky model as well as the popular structure function model [8], are quite successful in certain flow situations. Nevertheless, for some flows, e.g. wall-bounded shear flows, transitional or intermittent flows, *ad-hoc* remedies have to be used to get acceptable results (see section 3).

The application of spatial filters to separate the smaller resolved scales from the large-scale flow was, e.g., successfully used in the filtered structure function model [2]. Alternatively, high-pass-filtered (HPF) eddy-viscosity models have, recently and independently, been proposed in [21] and [19]. Here, the computation of the strain-rate relies on high-pass-filtered quantities $H * \overline{\mathbf{u}}$ with a suitable high-pass filter H. For the HPF Smagorinsky model proposed in [19], one gets

$$\nu_T^{\mathrm{HPF}} = (C_{S,\omega_c}^{\mathrm{HPF}} \Delta)^2 |S(H * \overline{\mathbf{u}})| \quad , \quad |S(H * \overline{\mathbf{u}})| = (2S_{ij}(H * \overline{\mathbf{u}})S_{ij}(H * \overline{\mathbf{u}}))^{1/2} \tag{7}$$

and

$$\tau_{ij} - \frac{\delta_{ij}}{3}\tau_{kk} = -2\nu_T^{\mathrm{HPF}} S_{ij}(H * \overline{\mathbf{u}}) . \tag{8}$$

Similarly, this approach can be applied in conjuction with the (filtered) structure function (SF/FSF) model [19]. These models can be used equally well for transitional and turbulent, incompressible and compressible flows [19, 15] (see results presented in section 3).

Another widely-used option is to resort to dynamic LES models, which were introduced in [3, 7]. These models, in particular the dynamic Smagorinsky model (DSMAG), attempt to adapt the (spatially and temporally varying) local model coefficient automatically to different flow situations, e.g. laminar, transitional, and turbulent flows. However, in general they do not correctly predict a laminar base flow. Moreover, averaging of the dynamic model coefficient in the homogeneous directions has usually to be employed for turbulent flows in order to minimize the occurrence of singularities.

Another recent modelling approach is the approximate deconvolution model (ADM) which was proposed in [16] for the LES of incompressible and compressible flows. The model is based on an approximate deconvolution of the

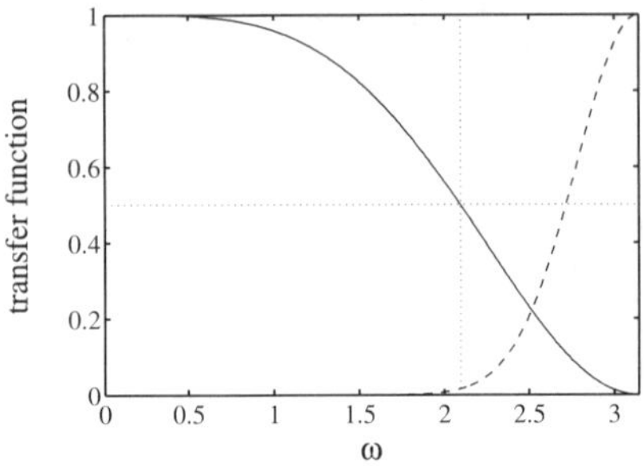

Fig. 1. Transfer functions of the different filter (equidistant grid) with $N = 5$ and $\omega_c = 2\pi/3$; ——— $\hat{G}(\omega)$, - - - - $\hat{H}_N(\omega) = 1 - \hat{Q}_N(\omega)\hat{G}(\omega)$.

filtered data by a truncated series expansion of the inverse filter. Furthermore, a relaxation term which acts only on the scales close to the numerical cut-off is used to model the interaction of the resolved scales with those not represented numerically.

The ADM SGS model to close equation 2 can be written as follows

$$\frac{\partial \tau_{ij}}{\partial x_j} = \frac{\partial \overline{u_j^\star u_i^\star}}{\partial x_j} - \frac{\partial \overline{u}_j \overline{u}_i}{\partial x_j} + \chi(I - Q_N * G) * \overline{u}_i \ . \tag{9}$$

Here, a star denotes the approximately deconvolved quantities $u_i^\star := Q_N * \overline{u}_i$. G is the graded discrete primary low-pass filter and Q_N its approximate inverse [16]

$$Q_N = \sum_{\nu=0}^{N} (I - G)^\nu = (I - G)^{N+1} \approx G^{-1} \ . \tag{10}$$

The definition of G is given in [17], see Fig. 1. The model coefficient χ of the relaxation term can be estimated from the instantaneous solution by a dynamic procedure [17].

SGS modeling using ADM has been applied to a variety of different flows showing very good agreement of the LES with filtered DNS results. Incompressible transitional and turbulent channel flow has been considered in [17, 13], compressible shock-turbulent-boundary-layer interaction in [18], and a transitional/turbulent rectangular jet in [10]. The applicability of ADM in low-order finite-volume CFD codes has been demonstrated in [20].

Based on equation 9, a variant of ADM can be formulated. The relaxation-term (RT) model simply uses

$$\frac{\partial \tau_{ij}}{\partial x_j} = \chi(I - Q_N * G) * \overline{u}_i \ . \tag{11}$$

For spectral discretization of the governing equations this model shows very good results for incompressible flows [13], even without the use of a dealiasing procedure (3/2-rule) for the convective terms [12].

3 LES of Plane Channel Flow Transition

3.1 LES of Laminar-Turbulent Transition

Transitional flows have been the subject of intense experimental and numerical research for many decades. One of the first well-resolved simulations to calculate from a basically laminar initial field through 3D transition and breakdown into fully developed turbulence was presented in the DNS work of [4], who considered fundamental K-type transition in plane Poiseuille flow. Comprehensive review articles on the numerical simulation of transition can be found in [5] and [11].

A SGS model suitable to simulate transition should naturally be able to deal equally well with laminar, transitional and turbulent flow states. The model should leave the laminar base flow unaffected and only be effective, in an appropriate way, when nonlinear interactions between the resolved modes and the non-resolved scales become important. The initial slow growth of the instability waves is mostly sufficiently resolved even on a coarse LES grid. Due to the possibly intermittent character of transitional flows (e.g. spatial simulations, bypass transition), spatial averaging of the model coefficients (a common practice with the dynamic Smagorinsky model) should be avoided. Furthermore, for the sake of generality, a three-dimensional formulation of the model is required. Ideally, the SGS model should not rely on an *ad-hoc* adaptation to a specific flow situation.

3.2 Simulation Results

As a LES test case, temporal incompressible K-type channel-flow transition was studied using a fully-spectral method with the same parameters as in [4].

The initial disturbances consist of a two-dimensional (stable) Tollmien-Schlichting (TS) wave and two superimposed weak oblique (stable) three-dimensional waves. The Reynolds number based on bulk velocity and channel half-width is $Re_b = 3333$, the box dimensions are $5.61 \times 2.99 \times 2$. For the statistically stationary results of fully-developed turbulent channel flow the data is statistically averaged over a non-dimensional time interval of $500-1000$ well after transition.

For the results presented in this section, an LES resolution of $32^2 \times 33$ grid points was chosen, whereas the reference DNS was performed on a $160^2 \times 161$ grid. Results are presented for the models listed in Table 1. Note that the resolution chosen for the LES is quite coarse [13]. The application of standard ADM [17] at this resolution failed due to a further restriction of the resolved scales by the filter G. However, at somewhat less coarse (especially in the wall-normal direction) resolution ADM works well [13]. On the other hand, the comparably good results obtained for the no-model calculation stems from the fact, that a very robust non-dissipative spectral discretization scheme with correction of aliasing errors has been used.

Table 1. Temporally and spatially averaged skin friction Reynolds number Re_τ obtained for the different simulations of fully developed turbulent channel flow.

$Re_b = 3333$	Re_τ	caption
RT $32^2 \times 33$	208.9	——
HPF-SMAG $32^2 \times 33$	208.8	- - - -
HPF-FSF $32^2 \times 33$	211.1	- - - -
DSMAG $32^2 \times 33$	195.1	—·—
no-model LES $32^2 \times 33$	220.0	——
fully-resolved DNS $160^2 \times 161$	208.2	
grid-filtered to $32^2 \times 33$	208.4	•

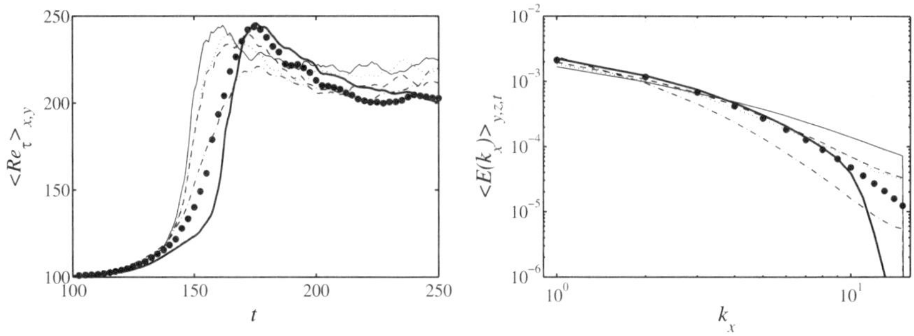

Fig. 2. *Left:* Evolution of Re_τ averaged in wall planes during the transitional phase. *Right:* Streamwise Fourier spectrum of the streamwise velocity component u for the turbulent phase, averaged in y, z and t. Line caption see Table 1.

In Figure 2, the evolution of the skin friction Reynolds number Re_τ is shown during the transition phase. As expected, the no-model calculation goes through transition earlier than the fully resolved calculation and the LES involving additional SGS dissipation. The two HPF models are quite similar to each other, whereas the RT model is clearly more dissipative during the inital stages of transition ($t \approx 150$). The dynamic Smagorinsky model shows an accurate prediction of the transitional process until $t \approx 170$, but fails to show the well-know overshoot of Re_τ of about 20% ($t \approx 175$).

The one-dimensional streamwise spectrum is also given in Figure 2. The strong influence of the relaxation term $-\chi(H * \overline{\mathbf{u}})$ can clearly be seen. Due to the unfiltered strain rate used in the calculation of τ_{ij} for the dynamic Smagorinksy model, its dissipative influence is visible throughout the whole wavenumber range.

Figure 3 presents some characteristics of the fully developed turbulent flow. The mean wall-normal velocity profile $u(z)$ and the Reynolds stresses are predicted accurately for the RT and HPF models; the dynamic Smagorinsky model is slightly too dissipative.

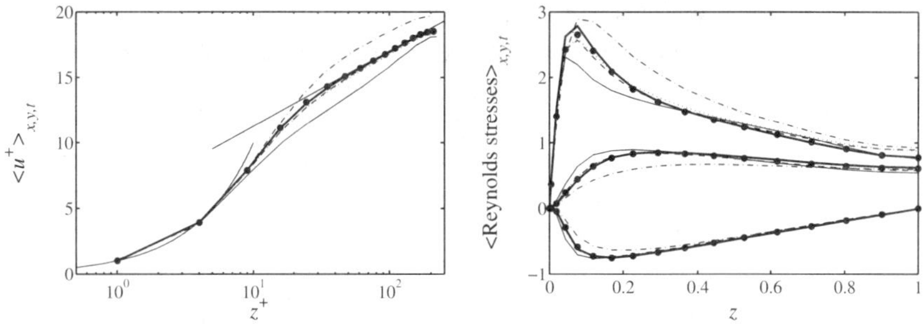

Fig. 3. *Left:* Averaged velocity profile u^+ scaled in wall units. *Right:* $(u_1'u_1')^{1/2}/u_\tau$, $(u_2'u_2')^{1/2}/u_\tau$ and $-u_1'u_3'/u_\tau^2$ in the fully turbulent case ($Re_b = 3333$). Line caption see Table 1.

4 Conclusions

In this short review, some recent LES techniques were presented and the basic difficulties of LES of transitional flows have been discussed. Results of several large-eddy simulations including a variant of the approximate deconvolution model (ADM), the relaxation-term (RT) model, the high-pass filtered eddy viscosity models, and the dynamic Smagorinsky model for transitional incompressible channel flow were presented and compared to no-model LES and fine-grid DNS calculations.

The results obtained indicate that it is well possible to simulate transitional wall-bounded flows on the basis of both ADM and HPF eddy-viscosity models with constant model coefficient. During the early stages of transition, the results of no-model LES calculations, which have sufficient resolution for this stage of flow development, are recovered. This confirms that the LES model is inactive there. During the rapid mean flow development, the model contributions are beginning to provide additional dissipation. The results demonstrate that a proper treatment of each spatial direction should be used in order to faithfully represent the relevant physical features such as the local gradients.

The LES models presented herein are completely dynamic in space and time such that no ad-hoc constants or adjustments are needed. This self-adaptation to the current local flow situation is very important for all types of transitional flows.

Acknowledgments

This work was supported by the Swiss National Science Foundation (SNF) and the Swiss Center for Scientific Computing (CSCS). Calculations were performed at CSCS.

References

1. J. A. Domaradzki and N. A. Adams. Direct modelling of subgrid scales of turbulence in large eddy simulations. *J. Turbulence*, 3:1–19, 2002.

2. F. Ducros, P. Comte, and M. Lesieur. Large-eddy simulation of transition to turbulence in a boundary layer developing spatially over a flat plate. *J. Fluid Mech.*, 326:1–36, 1996.
3. M. Germano, U. Piomelli, P. Moin, and W. H. Cabot. A dynamic subgrid-scale eddy viscosity model. *Phys. Fluids*, 3(7):1760–1765, 1991.
4. N. Gilbert and L. Kleiser. Near-wall phenomena in transition to turbulence. In S. J. Kline and N. H. Afgan, editors, *Near-Wall Turbulence – 1988 Zoran Zarić Memorial Conference*, pages 7–27. Hemisphere, 1990.
5. L. Kleiser and T. A. Zang. Numerical simulation of transition in wall-bounded shear flows. *Annu. Rev. Fluid Mech.*, 23:495–537, 1991.
6. M. Lesieur and O. Métais. New trends in large-eddy simulations of turbulence. *Annu. Rev. Fluid Mech.*, 28:45–82, 1996.
7. D. K. Lilly. A proposed modification of the Germano subgrid-scale closure method. *Phys. Fluids*, A 4(3):633–635, 1992.
8. O. Métais and M. Lesieur. Spectral large-eddy simulation of isotropic and stably stratified turbulence. *J. Fluid Mech.*, 239:157–194, 1992.
9. P. Moin and K. Mahesh. Direct numerical simulation: A tool in turbulence research. *Annu. Rev. Fluid Mech.*, 30:539–578, 1998.
10. B. Rembold, N. A. Adams, and L. Kleiser. Direct and large-eddy simulation of a transitional rectangular jet. In B.J. Geurts, R. Friedrich, and O. Métais, editors, *Direct and Large-Eddy Simulation IV*, pages 179–204. Kluwer, Dordrecht, The Netherlands, 2001.
11. D. Rempfer. Low-dimensional modeling and numerical simulation of transition in simple shear flows. *Annu. Rev. Fluid Mech.*, 35:229–265, 2003.
12. P. Schlatter, S. Stolz, and L. Kleiser. Relaxation-term models for LES of transitional/turbulent flows and the effect of aliasing errors. In *Proceedings of DLES-5, Munich, August 27-29, 2003*, 2003.
13. P. Schlatter, S. Stolz, and L. Kleiser. LES of transitional flows using the approximate deconvolution model. *Int. J. Heat Fluid Flow*, 2004. Accepted.
14. J. Smagorinsky. General circulation experiments with the primitive equations. *Mon. Weath. Rev.*, 91(3):99–164, 1963.
15. S. Stolz. High-pass filtered eddy-viscosity models for large-eddy simulations of compressible wall-bounded flows. In *Proceedings of HT-FED 2004, ASME Heat Transfer/Fluids Engineering Summer Conference*, 2004. Charlotte, North Carolina, USA, July 11-15.
16. S. Stolz and N. A. Adams. An approximate deconvolution procedure for large-eddy simulation. *Phys. Fluids*, 11(7):1699–1701, 1999.
17. S. Stolz, N. A. Adams, and L. Kleiser. An approximate deconvolution model for large-eddy simulation with application to incompressible wall-bounded flows. *Phys. Fluids*, 13(4):997–1015, 2001.
18. S. Stolz, N. A. Adams, and L. Kleiser. The approximate deconvolution model for large-eddy simulations of compressible flows and its application to shock-turbulent-boundary-layer interaction. *Phys. Fluids*, 13(10):2985–3001, 2001.
19. S. Stolz, P. Schlatter, D. Meyer, and L. Kleiser. High-pass filtered eddy-viscosity models for LES. In *Proceedings of DLES-5, Munich, August 27-29, 2003*, 2003.
20. R. von Kaenel, N. A. Adams, L. Kleiser, and J. B. Vos. The approximate deconvolution model for large-eddy simulation of compressible flows with finite volume schemes. *J. Fluids Eng.*, 125:375–381, 2003.
21. A. W. Vreman. The filtering analog of the variational multiscale method in large-eddy simulation. *Phys. Fluids*, 15(8):L61–L64, 2003.

Direct Numerical Simulations of Turbulent Rayleigh-Bénard Convection in Wide Cylinders

O. Shishkina and C. Wagner

DLR - Institute for Aerodynamics and Flow Technology, Göttingen
Olga.Chichkina@dlr.de, Claus.Wagner@dlr.de

It is well-known that heat transport in Rayleigh-Bénard convection, which can be expressed by the Nusselt number Nu, scales with the Rayleigh number Ra [1]-[2]. Chavanne *et al* compared numerous experimental results [3] and showed that the obtained scaling law Nu vs. Ra depends on the aspect ratio $a = H/R$ (where R denotes the radius and H - the height of the cylinder) and the Prandtl number $Pr = \nu/\kappa$. While most experiments were conducted in a cylindrical confinement for practical reasons, the majority of the so far performed numerical simulations were conducted for planar configurations [4]-[7]. Recently Verzicco and Camussi [8] presented results of Direct Numerical Simulations (DNS) in a slender cylinder with $a = 4$ for Ra up to 10^{11}. Our objective is to investigate turbulent Rayleigh-Bénard convection in wide cylinders with the aspect ratio $a < 1$ by means of DNS.

Governing Equations and Computational Outline

Using R, $R^{1/2}(\alpha g \Delta T)^{-1/2}$, $\alpha g R \Delta T \rho$ and $\Delta T = \hat{T}_1 - \hat{T}_2$ (difference between the temperatures $\hat{T}_1$ at the bottom and $\hat{T}_2$ at the top of the cylinder) as reference values for length, time, pressure and temperature, respectively, we receive the following dimensionless governing equations for the Rayleigh-Bénard problem

$$\mathbf{u}_t + \mathbf{u} \cdot \nabla \mathbf{u} + \nabla p = a^{3/2} Pr^{1/2} Ra^{-1/2} \nabla^2 \mathbf{u} + T\mathbf{z}, \quad \nabla \cdot \mathbf{u} = 0, \qquad (1)$$
$$T_t + \mathbf{u} \cdot \nabla T = a^{3/2} Pr^{-1/2} Ra^{-1/2} \nabla^2 T,$$

where $Ra = g\alpha H^3 \Delta T/(\nu\kappa)$, $\mathbf{u}$ and $\mathbf{u}_t$ are the velocity field and its time derivative, T is the temperature, p is the pressure, ν is the kinematic viscosity, κ is the thermal diffusivity, g is the gravitational acceleration, ρ is the density and α is the termal expansion coefficient. On all boundaries of the cylindrical container $\mathbf{u} = \mathbf{0}$, while T varies from $+0.5$ at the bottom to -0.5 at the top of the cylinder. Further, an adiabatic vertical wall is prescribed by $\partial T/\partial r = 0$.

The DNS were performed on staggered grids with $128 \cdot 512 \cdot 192$ nodes (for higher Ra) and $96 \cdot 256 \cdot 128$ nodes (for lower Ra) in (z, φ, r)-directions of the cylindrical coordinate sytem. A randomly disturbed linear temperature distribution $T = 0.5 - z/a$ and $\mathbf{u} = \mathbf{0}$ served as the initial conditions. For solving (1) we developed a finite volume method, which uses fourth order accurate central differences in spatial directions and a hybrid explicit/semi-implicit time stepping scheme as described in [9].

Instantaneous Snapshots of the Velocity Field

In Fig.1 instantaneuos snapshots of the axial velocity are depicted for $z = a/2$ and different Ra and a. Positive velocity values (the flow rises) correspond to light and negative values - to dark areas. For $Ra = 10^5$, $a = 0.2$, elongated structures can be observed. If Ra is increased to 10^7, fine flow structures evolve on top of the large structure leading to a more corrugated large scale flow. A further increase of Ra ($a = 0.4$) leads to a single large flow structure. The latter is characterized by warm fluid rising in the center and cold fluid descending close to the cylinder sidewall.

$Ra = 10^5$, $a = 0.2$ $Ra = 10^7$, $a = 0.2$ $Ra = 10^8$, $a = 0.4$

Fig. 1. Instantaneuos snapshots of the axial velocity u_z for $z = a/2$

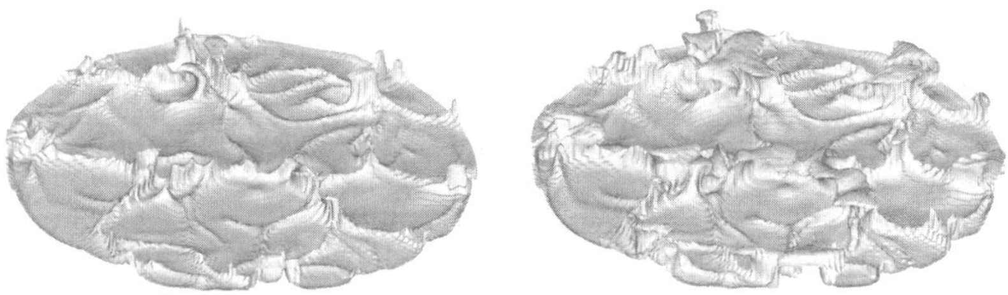

$T = 0.25$ $T = 0.15$

Fig. 2. The isotherms of the solution for $Ra = 10^5$, $a = 0.2$

In Fig.2 the side views of isosurfaces $T = 0.15$ and $T = 0.25$ for the case $Ra = 10^5$, $a = 0.2$ are presented. It is observed that the warm fluid ($T > 0$) reaches the upper cylinder wall in burr-like structures. Vice versa the cold fluid ($T < 0$) reaches the vicinity of the bottom wall in the cracks-like stucture (not shown).

Mean Flow Characteristics

Time averaging was performed for more than 500 dimensionless time units until no significant change of the absolut maximum velocity value was observed. From contours of the mean (time averaged) axial velocity, presented in Fig.3 for central cross-sections and different Ra and a, we conclude a growth in size of the mean flow structures with increasing Ra. Additionally averaging the mean axial flow fields in azimuthal direction we receive two-dimensional flow fields, which are presented in Fig.4. For all cases the mean flow rises in the center of the cylinder.

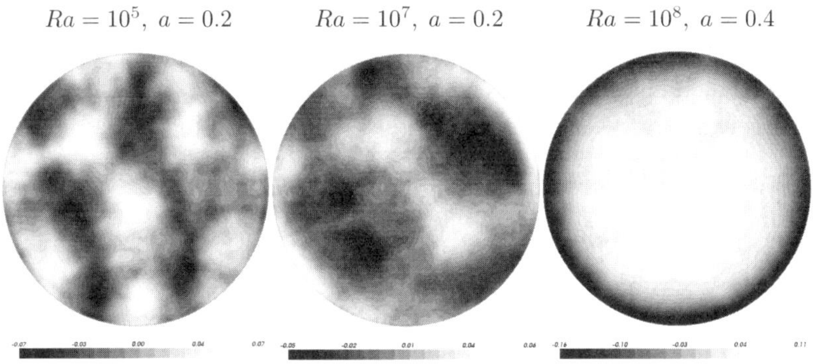

Fig. 3. Contours of time averaged mean axial velocity component u_z, $z = a/2$

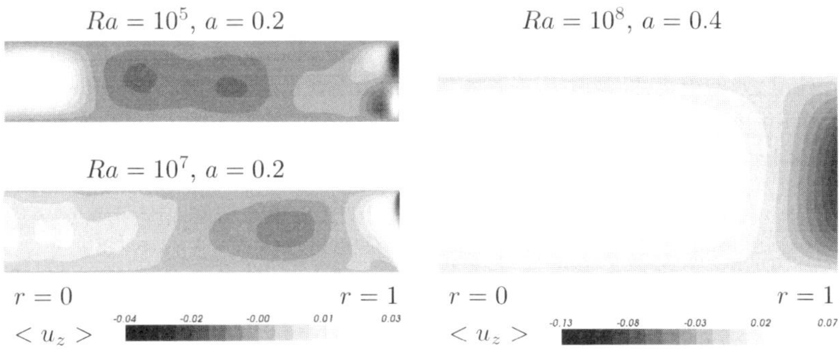

Fig. 4. Mean axial velocity $< u_z >$ averaged in time and in φ-direction

Finally, the scaling behaviour of $Nu = \langle u_z T \rangle H u_b / \kappa - a \partial < T > / \partial z$ with Ra, where $< \cdot >$ denotes the average over an arbitary horizontal cross section of the cylinder, was investigated. Evaluation of the DNS data obtained for $a = 0.2$ led to the following Nu-numbers: $Nu = 4.29$ for $Ra = 10^5$, $Nu = 8.81$ for $Ra = 10^6$ and $Nu = 18.13$ for $Ra = 10^7$, which fit the scaling law $Nu = 0.117 Ra^{0.313}$.

Spatial and Temporal Resolution of the Solution

The mean mesh size $h = \max\left(\Delta z \cdot r\Delta\varphi \cdot \Delta r\right)^{1/3}$ must be of the same order as the Kolmogorov's scale $\eta = aPr^{1/2}(Nu-1)^{-1/4}Ra^{-1/4}$, namely $h = O(\pi\eta)$, in order to resolve all relevant turbulent length scales in a DNS. For $Ra = 10^5$, $a = 0.2$ this leads to $\pi\eta = 2.19 \cdot 10^{-2}$ and for $Ra = 10^7$, $a = 0.2$ we obtain $\pi\eta = 4.53 \cdot 10^{-3}$. On the other hand, for the mesh with $96 \cdot 256 \cdot 128$ grid points in z-, φ-, r-directions, respectively, which we used in the DNS, $h = 6.49 \cdot 10^{-3}$. Therefore, it is concluded that the spatial resolution used in the DNS is fine enough for the cases $Ra \leq 10^7$.

The temporal resolution requirements are the following: 1) the time step must be smaller than the Batchelor scale $\eta_T = a^{1/2}Pr^{1/2}(Nu-1)^{-1/2}$ and 2) it has to be small enough to guarantee the numerical stability. In [10] the sufficient condition for the von Neumann stability of the explicit time integration scheme based on central differences of any even order was presented. For the fourth order accurate scheme the critical time step equals

$$\Delta t < \left[\max_{z,\varphi,r}\left\{\frac{3}{2}\left[\frac{|u_z|}{\Delta z} + \frac{|u_\varphi|}{r\Delta\varphi} + \frac{|u_r|}{\Delta r}\right] + \frac{16\sqrt{a^3 Pr}}{3\sqrt{Ra}}\left[\frac{1}{\Delta z^2} + \frac{1}{(r\Delta\varphi)^2} + \frac{1}{\Delta r^2}\right]\right\}\right]^{-1},$$

which is additionally multiplied by safety factor 0.5 to avoid nonlinear instabilities. From this we obtained the time step $\Delta t_{DNS} = \min\{\eta_T; 0.5\Delta t\} = 0.5\Delta t$.

Conclusions and Outlook

DNS of turbulent Rayleigh-Bénard convection in cylindrical containers with aspect ratios $a = 0.4$, $a = 0.2$ were performed for $10^5 < Ra < 10^8$ on cylindrical grids with up to 13 million grid points. It was shown that the spatial and temporal resolutions were sufficient to resolve all relevant turbulent scales. The instantaneous flow fields reveal large elongated flow structures, which develop in the considered moderate Ra regime. Additionally, large flow structures grow with increasing Ra. More specifically, an increase of Ra from 10^7 to 10^8 leads to a transition from cell-like turbulent flow structures to a single circulation, which rises in the center and descends close to the sidewall.

References

1. Siggia ED (1994) Ann Rev Fluid Mech 26:137–168
2. Grossmann S, Lohse D (2000) J Fluid Mech 407:27–56
3. Chavanne X, Chillà F, Chabaud B, Castaing B, Hebral B (2001) Phys Fluids 13:1300–1320
4. Kimmel SJ, Domaradzki JA (2000) Phys Fluids 12:169-184
5. Grötzbach G (1982) J Fluid Mech 119:27-53
6. Kerr RM (1996) J Fluid Mech 310:139-179
7. Hartlep T, Tilgner A, Busse FH (2003) Phys Rev Lett 91(6):1-6
8. Verzicco R, Camussi R (2002) J Fluid Mech, preprint
9. Shishkina O, Wagner C (2004) Computers and Fluids, submitted
10. Shishkina O, Wagner C (2004) Appl Numer Anal and Comput Math, 1: 323-334

A Projective Similarity/Eddy-Viscosity Model for Large-Eddy Simulation

Roel Verstappen

Research Institute of Mathematics and Computer Science
University of Groningen, P.O.Box 800, 9700 AV Groningen, The Netherlands
verstappen@math.rug.nl

1 Introduction

Since most turbulent flows possess far more eddies than can be computed from the Navier-Stokes equations, an approximate description of the dynamics of the 'large-eddies' is sought in which the 'small-eddies' need not be computed explicitly. In large-eddy simulation (LES), this is achieved by applying a frequency low-pass filter to the equations; see e.g. [1]. The interaction between the large (filtered) and small (residual) eddies is then represented by the commutator of the filter and the nonlinear term in the Navier-Stokes equations. Hence, this commutator has to be modeled in terms of the filtered velocity to obtain the intended, approximate, large-eddy dynamics. For this, one usually resorts to an eddy-viscosity model cf. [2]-[3], a (scale) similarity model cf. [4]-[5], or a mix thereof cf. [6]. Similarity models have the proper mathematical structure. Additionally, they correlate well with the real commutator. Yet, their leading term has directions of negative diffusion [7]-[8]. In this paper, we propose to stabilize similarity models by projecting them onto an eddy-viscosity model. The projection eliminates the dynamically unstable part and results in a self-calibrating eddy-viscosity. In comparison with mixed models, we do not add a dissipative term to stabilize the similarity model, but instead we remove the dynamically unstable art. The resulting projective similarity/eddy-viscosity model is successfully tested for a turbulent channel flow at $\mathrm{Re}_\tau = 2520$ (based on the friction velocity and channel half-width).

2 Spatial Filter

We consider the elliptic, differential filter [9]

$$\bar{u} = \mathcal{F}u = u + \alpha^2 \nabla^2 u, \tag{1}$$

where α parameterizes the filter-length. The boundary conditions that supplement the Navier-Stokes equations are applied to the filter too. This filter is generic in the sense that any symmetric convolution filter can be approximated by (1), where the error is of the order α^4. Additionally, it has been shown in [10] that the approximate inverse of (1),

$$u \approx \tilde{u} = \tilde{\mathcal{F}}^{-1}\bar{u} = (1 - \alpha^2\nabla^2)\bar{u}, \tag{2}$$

forms the essence of the recently proposed 'alpha-model'.

The commutator of the filter (1) and the Navier-Stokes operator is given by $\nabla \cdot \tau$, where the subfilter stress τ depends upon the velocity-gradient:

$$\tau_{ij}(\nabla u) = 2\alpha^2\nabla u_i \cdot \nabla u_j - \alpha^4\nabla^2 u_i \nabla^2 u_j. \tag{3}$$

3 Projective Similarity/Eddy-Viscosity Model

Similarity models are based upon an approximate defiltering procedure. With the help of the approximate defilter given by (2), we can model the subfilter stress τ_{ij} by replacing the velocity u in (3) by the right-hand side of (2). The resulting similarity model $\tau_{ij}(\nabla u) \approx \tau_{ij}(\nabla\tilde{\mathcal{F}}^{-1}\bar{u})$ possesses the correct mathematical structure; particularly, it satisfies all properties of a commutator. Additionally, the correlation between the approximation $\tau_{ij}(\nabla\tilde{u})$ and $\tau_{ij}(\nabla u)$ is generally strong, typically between 0.6 and 0.9. Yet, this model is not unconditionally stable as the leading term of $\tau_{ij}(\nabla\tilde{u})$ has directions of negative dissipation. Therefore, we propose to remove the dynamically unstable part of $\tau_{ij}(\nabla\tilde{u})$ by means of a projection onto an eddy-viscosity model of the form

$$-\tau_{ij}(\nabla u) + \tfrac{1}{3}\delta_{ij}\tau_{kk}(\nabla u) \approx \nu(\partial_j\bar{u}_i + \partial_i\bar{u}_j), \tag{4}$$

where the isotropic part $\frac{1}{3}\delta_{ij}\tau_{kk}(\nabla u)$ need not be modeled, as it can be incorporated into the pressure. The projection results into a self-calibrating eddy-viscosity $\nu(x,t)$ which is computed such that the best approximation of $-\tau_{ij}(\nabla\tilde{u}) + \frac{1}{3}\delta_{ij}\tau_{kk}(\nabla\tilde{u})$ is obtained in least-square sense,

$$\min \int e_{ij}e_{ij}\mathrm{d}V, \tag{5}$$

where integral extends over the entire flow domain, the residuals are $e_{ij} = \tau_{ij}(\nabla\tilde{u}) - \frac{1}{3}\delta_{ij}\tau_{kk}(\nabla\tilde{u}) - \nu(\partial_j\bar{u}_i + \partial_i\bar{u}_j)$ and the minimum is computed (with respect to ν) subject to the stability constraint $\nu + 1/\mathrm{Re} > 0$. The solution of this constrained variational problem reads

$$\nu = \frac{(\tau_{ij}(\nabla\tilde{u}) - \frac{1}{3}\delta_{ij}\tau_{kk}(\nabla\tilde{u}))(\partial_j\bar{u}_i + \partial_i\bar{u}_j)}{(\partial_n\bar{u}_m + \partial_m\bar{u}_n)^2}, \tag{6}$$

if the right-hand side is larger than $-1/\mathrm{Re}$; and $\nu = -1/\mathrm{Re}$ otherwise.

In summary, the projective similarity/eddy-viscosity model is given by (4), where the eddy-viscosity ν is computed according to (6), with $\tilde{u}$ as in (2) and τ_{ij} as in (3).

4 An a Posteriori Test: Turbulent Channel Flow

As a first step in application of the proposed model, it is tested for a turbulent channel flow at $\mathrm{Re}_\tau = 2520$ by comparing the results with those of Comte-Bellots wind tunnel experiment [11]. Besides we will compare with experiments by Wei and Willmarth [12], which were performed at lower Re_τ (1025-1650), and since making good near-wall measurements is difficult, we also make a comparison with a direct numerical simulation (DNS) at $\mathrm{Re}_\tau = 590$ [13]. Obviously, the comparison with DNS is to be restricted to the direct vicinity of the wall, where Reynolds-number effects can be properly scaled.

As usual, the flow is assumed to be periodic in the stream- and spanwise direction. The computational grid consists of 128 streamwise points, 64 spanwise points and 300 points between the channel walls. All LES-results are approximately defiltered by means of (2) in order to compare them directly with the available experimental data. Details of the computational procedure are discussed elsewhere [14]; here we focus on the principal results.

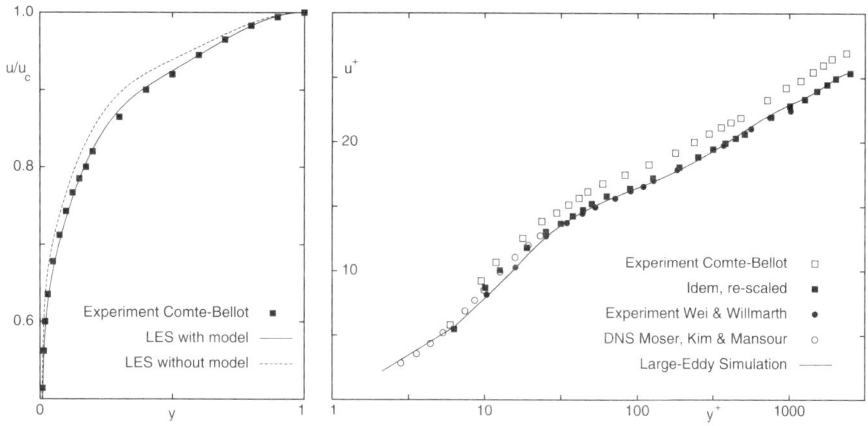

Fig. 1. Mean streamwise velocity. In the left-hand side the velocity is normalized by the centre-line velocity. The right-hand side figure shows u^+ as function of y^+.

The least to be expected from a LES is a good prediction of the mean flow. As can be seen in Fig. 1 (left) the present LES satisfies that minimal requirement: without any models the prediction of the mean flow worsens significantly. With model, the agreement with the data of Comte-Bellot is good. Yet, the friction velocities u_τ differ. Comte-Bellot deduced $u_\tau = 0.0416$. We have $u_\tau = 0.0442$, which is in good agreement with Dean's result $u_\tau = 0.0445$. Therefore, we have rescaled the Comte-Bellots mean-velocity profile with the help of our u_τ. After this rescaling, the result of Comte-Bellot shows an excellent agreement with the data in [12]-[13] and with the present LES.

As can be seen in Fig. 2, the turbulent intensities agree well, except for the spanwise fluctuations, which agree fairly: the spanwise turbulence intensity of

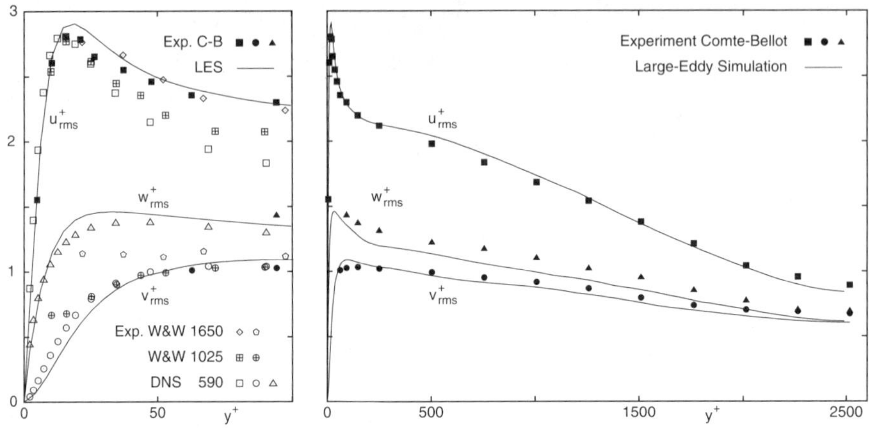

Fig. 2. The root-mean-square of the fluctuating velocity at $Re_\tau = 2520$. Here, the results by Comte-Bellot are rescaled, like in Fig. 1. Near the wall (left-hand figure) two experiments by Wei and Willmarth [12] are shown, namely at $Re_\tau = 1650$ and $Re_\tau = 1025$. The DNS [13] has been performed at $Re_\tau = 590$. Note: for $y^+ > 30$ the comparison with low-Re_τ data does not hold due to Reynolds-number effects.

Comte-Bellot is consistently higher than that of the LES. Near the wall there also exists a good agreement between the streamwise intensity measured by Wei and Willmarth ($Re_\tau = 1650$) and the present result. In summary, good agreement with previously reported experimental results is observed.

References

1. Sagaut P (2001) Large eddy simulation for incompressible flows. Springer, Berlin
2. Smagorinsky J (1963) Month Weath Rev 91(3):99–165
3. Germano M (1992) J Fluid Mech 238:325–336
4. Bardina J, Ferziger JH, Reynolds WC (1984) Improved turbulence models based on LES of homogeneous incompressible turbulent flows. Report TF-19, Stanford
5. Stolz S, Adams NA (1999) Phys Fluids 11(7):1699–1701
6. Clark RA, Ferziger JH, Reynolds WC (1979) J Fluid Mech 91:1–16
7. Leonard A (1997) Large-eddy simulation of chaotic convection and beyond. AIAA Paper 97-0204
8. Vreman B (1995) Direct and large-eddy simulation of the compressible turbulent mixing layer. PhD thesis, University of Twente, Enschede
9. Germano M (1986) Phys Fluids 29(6):1755–1758
10. Montgomery DC, Pouquet A (2002) Phys Fluids 14(9):3365–3366
11. Comte-Bellot G (1963) Turbulent flow between two parallel walls. PhD thesis, University of Grenoble
12. Wei T, Willmarth WW (1989) J Fluid Mech 204:57–95
13. Moser RD, Kim J, Mansour NN (1999) Phys Fluids 11:943–945
14. Verstappen RWCP, Veldman AEP (2003) J Comp Phys 187:343–368

Passive Scalar Transport in Turbulent Supersonic Channel Flow

Holger Foysi and Rainer Friedrich

Fachgebiet Strömungsmechanik, TU München
Boltzmannstr. 15, D-85748 Garching
holger@flm.mw.tu-muenchen.de

Summary. Direct numerical simulations (DNS) of turbulent supersonic channel flow of air at Reynolds numbers ranging from $Re_\tau = 180$ to 560 and Mach numbers ranging from $M = 0.3$ to 3.0 have been performed. The DNS data are used to explain the reduction of the pressure-correlation terms due to compressibility, using a Green's function approach.

Compressibility Effects in Supersonic Channel Flow

Tables 1 and 2 show the parameters of the performed simulations. A homogeneous body force has been imposed on the momentum equation in streamwise direction to drive the mean flow and to allow for periodic pressure boundary conditions. The size of the domain has been adopted to be comparable to that of Coleman *et al.* [1] and Moser *et al.* [5]. The walls are cooled and kept at constant temperature to achieve supersonic flow, the passive scalar concentration is kept fixed at both walls and of opposite sign, as shown for the Favre averaged mean profiles in figure 1a. This corresponds to the introduction of the scalar on one wall and its removal from the other. The Prandtl and Schmidt numbers are 0.71 and 1, respectively.

Table 1. Simulation parameters (1/2)

Case	M	Re_τ	Re	$\frac{L_{x1}}{h}$	$\frac{L_{x2}}{h}$	$\frac{L_{x3}}{h}$	N_{x1}	N_{x2}	N_{x3}
M0.3	0.3	181	2850	9.6	6	2	192	160	129
M1.5	1.5	221	3000	4π	$4\pi/3$	2	192	128	151
M2.5	2.5	455	5000	2π	$2\pi/3$	2	256	128	201
M3.0	3.0	560	6000	2π	$2\pi/3$	2	256	128	221

Green's Function Analysis of the Pressure Correlation Terms

As recognized earlier by Huang *et al.* [4], the most important compressibility effect in supersonic channel flow is due to mean property variation in the

Table 2. Simulation parameters (2/2)

Case	Δx_1^+	$\Delta x_{2\,\mathrm{min}}^+$	$\Delta x_{2\,\mathrm{max}}^+$	Δx_3^+	τ_w	$\rho_w,$	mu_w	T_w
M0.3	14.46	1.12	4.19	7.23	7.555	1.210	0.0502	293
M1.5	14.46	0.85	5.02	7.23	30.993	0.023	1.91e-5	500
M2.5	11.16	1.23	7.46	7.44	79.710	0.035	1.91e-5	500
M3.0	13.37	0.89	10.22	8.91	110.156	0.042	1.91e-5	500

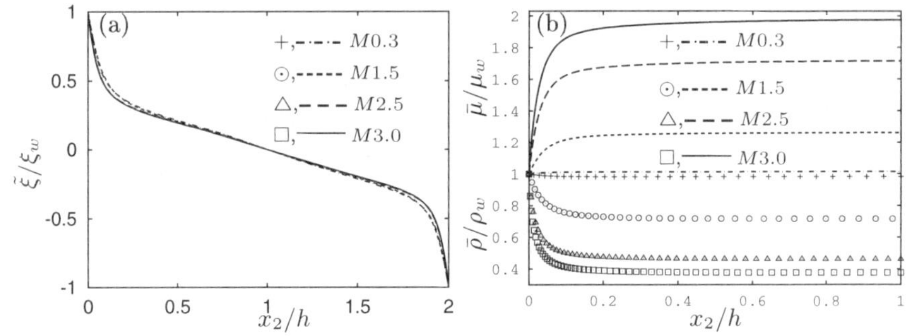

Fig. 1. Favre averaged scalar profile. (b) Variation of mean density (symbols) and mean viscosity (lines).

near wall region, as shown in figure 1b for cases M0.3 to M3.0. Huang *et al.* [4] therefore suggested to replace in the compressible cases the wall scaling usually adopted in incompressible channel flow by the so called semi-local scaling, which uses the local values of density and viscosity. Despite its success in scaling the mean velocity profiles, there are still discrepancies concerning the various terms in the Reynolds stress and scalar flux budgets. As shown in [2] and [3], the pressure correlation terms, for various Reynolds and Mach number cases, show large differences when scaled in inner variables, although a good collapse of the data for the streamwise production, dissipation and diffusion terms is observed. These observations and the fact that the pressure-strain correlation plays a key role in changing the turbulent stresses and their associated anisotropy and is, like the pressure-scalar-gradient-correlation difficult to model, make a more fundamental investigation necessary. To get further insight into the pressure-scalar-gradient correlation and pressure-strain correlation,

$$\Pi_i^\xi = \overline{p'\partial\xi''/\partial x_i} \quad \text{and} \quad \Pi_{ij}^u = \overline{p's'_{ij}} = \overline{p'(\partial u_i''/\partial x_j + \partial u_j''/\partial x_i)},$$

a starting point is to examine the compressible Poisson equation for the pressure fluctuations $\nabla^2 p' = \bar{\rho} f$, which reads

$$\nabla^2 p' = -\bar{\rho}(u_i''u_j'' - \overline{u_i''u_j''})_{,ij} - 2\bar{\rho}\widetilde{u}_{1,2}u''_{2,1} + \sigma'_{ij,ij} - 2\bar{\rho}_{,2}(u_2''u_j'' - \overline{u_2''u_j''})_{,j}$$
$$-\bar{\rho}_{,22}(u_2''^2 - \overline{u_2''^2}) - 2\widetilde{u}_{1,2}(\rho' u_2'')_{,1} - (\rho' u_i''u_j'' - \overline{\rho' u_i''u_j''})_{,ij} - D_{tt}\rho'.$$

The first two terms on the rhs are labeled A1 (nonlinear fluctuation), A2 (mean shear) and occur in a similar way in the incompressible case. In compressible flow A3 (viscous stress), B1 (density gradient), B2 (second density gradient), and three terms involving the density fluctuation, labeled C1, C2 and C3, exist. One important assumption in the following analysis is to neglect the density fluctuation terms, which are small ([3]), and normally lead to a convected wave equation for the pressure. The pressure-correlation terms can now be expressed as integrals over the Green's function G of equation (1)

$$\Pi^u_{ij}(y) = \int_{-1}^{1} \bar{\rho}(y')\,\overline{G * f'(x_1, y, x_3; y')s'_{ij}}\,dy' + \overline{B's'_{ij}} \tag{1}$$

$$\Pi^\xi_i(y) = \int_{-1}^{1} \bar{\rho}(y')\,\overline{G * f'(x_1, y, x_3; y')\frac{\partial \xi''}{\partial x_i}}\,dy' + \overline{B'\frac{\partial \xi''}{\partial x_i}} \tag{2}$$

with G and B being the back-transform of the Green's function in Fourier space ($k = \sqrt{k_1^2 + k_3^2}$, y is the wall normal coordinate with origin at the channel centerline)

$$\hat{G}(k, y; y') = -\frac{\cosh[k(y'-1)]\,\cosh[k(y+1)]}{k \sinh 2k} \quad , \quad y < y'$$

$$\hat{G}(k, y; y') = -\frac{\cosh[k(y'+1)]\,\cosh[k(y-1)]}{k \sinh 2k} \quad , \quad y > y'$$

and of the boundary term $\hat{B}$,

$$\hat{B} = \frac{\partial\hat{p}/\partial y\big|_{y=1} \cosh(k(1+y)) - \partial\hat{p}/\partial y\big|_{y=-1} \cosh(k(1-y))}{k \sinh 2k},$$

respectively (see [3] for details). Figures 2a and 3a show a comparison of the DNS results for the pressure-correlations and the results obtained with equations (1) and (2), for cases M0.3 and M1.5 (with similar Re_τ). Both indicate an excellent agreement, justifying our ansatz for the pressure fluctuations. They show furthermore, that the acoustic mode doesn't contribute very much to the pressure-correlation terms. In Figures 2b and 3b the influence of the different source terms of equation (1) is shown, normalized by $\tau_w u_{av}/h$ and $\chi u_{av}/h$, respectively (χ is the wall scalar flux). The nonlinear term A1 is clearly giving the largest contribution, followed by the source term B1. To mimic the effect of incompressibility, the mean density occurring in the source terms is replaced by its wall value, indicated by the square symbol. The resulting curves then clearly approach the incompressible solution (triangles) and suggest, that the decrease in magnitude of the peak pressure-correlation amplitudes observed in the compressible cases is caused to a large extent by the mean density decrease.

Conclusions

Several DNS of passive scalar transport in turbulent supersonic channel flow have been performed. An analysis using the Green's function of the pressure-

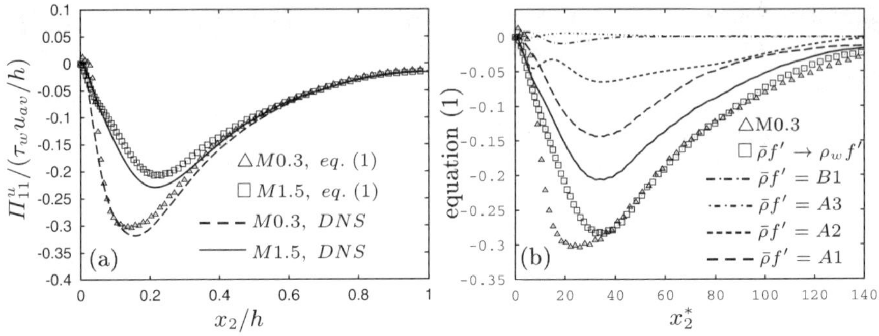

Fig. 2. (a) Comparison of the DNS data and Eq. (1) for the pressure-strain correlation. (b) Contribution of different source terms, $\bar{\rho} f'$, on the rhs of the pressure Poisson equation to Eq. (1) for Π^u_{11} in case M1.5.

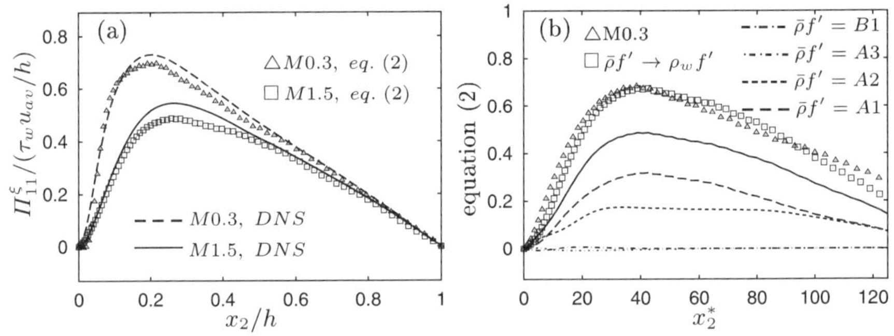

Fig. 3. (a) Comparison of the DNS data and Eq. (2) for the pressure-scalar-gradient correlation. (b) Contribution of different source terms, $\bar{\rho} f'$, on the rhs of the pressure Poisson equation to Eq. (2) for Π^ξ_1 in case M1.5.

fluctuation Poisson equation revealed, that the mean density decrease accounts for most of the reduction of the pressure correlation terms in the compressible case. Furthermore, it is seen, that the acoustic mode inherent in the density fluctuations of the Poisson equation does not contribute very much to the pressure correlation terms. To be able to better distinguish between Mach and Reynolds number effects, simulations at higher Reynolds number are currently performed.

Acknowledgments

The work was financially supported by the Deutsche Forschungsgemeinschaft under grant no. Fr478/20-1.

References

1. G.N. Coleman, J. Kim, and R.D. Moser. Turbulent supersonic isothermal-wall channel flow. *J. Fluid Mech.*, 305:159–183, 1995.
2. H. Foysi and R. Friedrich. DNS of Passive Scalar Transport in Turbulent Supersonic Channel Flow. *Proceedings, Third International Symposium on Turbulence and Shear Flow Phenomena, Sendai, Japan, 2003,*, Vol. III:1108–1116, 2003.
3. H. Foysi, S. Sarkar, and R. Friedrich. Compressibility Effects and Turbulence Scalings in Supersonic Channel Flow. *submitted to J. Fluid Mech.*, 2003.
4. P.G. Huang, G.N. Coleman, and P. Bradshaw. Compressible turbulent channel flows: DNS results and modelling. *J. Fluid Mech.*, 305:185–218, 1995.
5. R.D. Moser, J. Kim, and N. N. Mansour. Direct numerical simulation of turbulent channel flow up to $Re_\tau = 590$. *Phys. Fluids*, 9:943–945, 1999.

SPRINGER PROCEEDINGS IN PHYSICS

50 **Magnetic Properties of Low-Dimensional Systems II**
New Developments
Editors: L.M. Falicov, F. Mejía-Lira, and J.L. Morán-López

51 **The Physics and Chemistry of Organic Superconductors**
Editors: G. Saito and S. Kagoshima

52 **Dynamics and Patterns in Complex Fluids**
New Aspects of the Physics–Chemistry Interface
Editors: A. Onuki and K. Kawasaki

53 **Computer Simulation Studies in Condensed-Matter Physics III**
Editors: D.P. Landau, K.K. Mon, and H.-B. Schüttler

54 **Polycrystalline Semiconducturs II**
Editors: J.H. Werner and H.P. Strunk

55 **Nonlinear Dynamics and Quantum Phenomena in Optical Systems**
Editors: R. Vilaseca and R. Corbalán

56 **Amorphous and Crystalline Silicon Carbide III, and Other Group IV–IV Materials**
Editors: G.L. Harris, M.G. Spencer, and C.Y. Yang

57 **Evolutionaly Trends in the Physical Sciences**
Editors: M. Suzuki and R. Kubo

58 **New Trends in Nuclear Collective Dynamics**
Editors: Y. Abe, H. Horiuchi, and K. Matsuyanagi

59 **Exotic Atoms in Condensed Matter**
Editors: G. Benedek and H. Schneuwly

60 **The Physics and Chemistry of Oxide Superconductors**
Editors: Y. Iye and H. Yasuoka

61 **Surface X-Ray and Neutron Scattering**
Editors: H. Zabel and I.K. Robinson

62 **Surface Science**
Lectures on Basic Concepts and Applications
Editors: F.A. Ponce and M. Cardona

63 **Coherent Raman Spectroscopy**
Recent Advances
Editors: G. Marowsky and V.V. Smirnov

64 **Superconducting Devices and Their Applications**
Editors: H. Koch and H. Lübbing

65 **Present and Future of High-Energy Physics**
Editors: K.-I. Aoki and M. Kobayashi

66 **The Structure and Conformation of Amphiphilic Membranes**
Editors: R. Lipowsky, D. Richter, and K. Kremer

67 **Nonlinearity with Disorder**
Editors: F. Abdullaev, A.R. Bishop, and S. Pnevmatikos

68 **Time-Resolved Vibrational Spectroscopy V**
Editor: H. Takahashi

69 **Evolution of Dynamical Structures in Complex Systems**
Editors: R. Friedrich and A. Wunderlin

70 **Computational Approaches in Condensed-Matter Physics**
Editors: S. Miyashita, M. Imada, and H. Takayama

71 **Amorphous and Crystalline Silicon Carbide IV**
Editors: C.Y. Yang, M.M. Rahman, and G.L. Harris

72 **Computer Simulation Studies in Condensed-Matter Physics IV**
Editors: D.P. Landau, K.K. Mon, and H.-B. Schüttler

73 **Surface Science**
Principles and Applications
Editors: R.F. Howe, R.N: Lamb, and K. Wandelt

74 **Time-Resolved Vibrational Spectroscopy VI**
Editors: A. Lau, F. Siebert, and W. Werncke

75 **Computer Simulation Studies in Condensed-Matter Physics V**
Editors: D.P. Landau, K.K. Mon, and H.-B. Schüttler

76 **Computer Simulation Studies in Condensed-Matter Physics VI**
Editors: D.P. Landau, K.K. Mon, and H.-B. Schüttler

77 **Quantum Optics VI**
Editors: D.F. Walls and J.D. Harvey

78 **Computer Simulation Studies in Condensed-Matter Physics VII**
Editors: D.P. Landau, K.K. Mon, and H.-B. Schüttler